奇异花园
unique gardens

法国亦西文化(ICI CONSULTANTS) 筹划
(法) 苏菲·巴尔波(Sophie BARBAUX) 编著

Direction 总企划: Chia-Ling CHIEN 简嘉玲
Editorial Coordination 协调编辑: Nicolas BRIZAULT 尼古拉·布里左
English Translation 英文翻译: Kirsten SHEPARD 柯尔斯顿·薛帕尔德
Chinese Translation 中文翻译: Xue-Mei SHAO 邵雪梅 & Mei-Wen WANG 王美文
Chinese proofreading 中文校阅: Chia-Ling CHIEN 简嘉玲
Graphic Design 版式设计: Karine de La MAISON 卡琳·德拉梅宗
Layout 排版: Morgane SABOURET 茉尔甘·萨布黑

奇异花园
unique gardens

辽宁科学技术出版社

目录
CONTENTS

	前言 PREFACE	006
01	植物之歌 VEGETATION	012
02	协奏三曲 ACCOMPLICES	074
03	居民造景 INHABITANT LANDSCAPERS	144
04	馥丽小筑 FOLLIES	230
05	实景艺术 LIFE-SIZE ARTS	264
	附录 ANNEX	310

前言
preface

柑橘的神奇力量 /
orange power - 2008
Roberto Capecci, Marco Anonini &
Raffaela Sini, Land - I archicoltura (意大利)
Ponte de Lima International Garden Festival
(葡萄牙)
Photos ©: Land - I archicoltura
p.2

摇篮曲花园 /
lullaby garden - 2004
Cao | Perrot (美国&法国)
Cornerstone Gardens Festival, Sonoma (美国)
Photos ©: Stephen Jerrome
p.4

克拉姆溪迷宫 /
crum creek meander - 2014
Stacy Levy (美国)
Swarthmore College (美国)
Photos ©: Lia Roggio-Smith
pp.6-7

"如果人的一生都必须遵循已知事物而行事、自我局限于一小部分他所认识的现象——不论是通过教育或祖传意识所获得的，并且与其他人彼此联结、建立关系网络，这些纯功能性的组织网迟早会成为沉闷的陷阱，一个缺乏欲望的监牢。而他也将在其中被捆绑于逻辑这块黑硬面包与滞碍死水之间，直至腐朽。"

"If during his whole life man had to remain in the known, to keep limited to the small group of phenomena that he knows, by education and inheritance, to connect between them and form a network of relationships, this purely utilitarian network could not help but becoming a trap of boredom, a prison without desires in which he would be condemned to rot bound between the black bread and the stagnant water of logic."

米歇尔·莱里斯 Michel Leiris
试论西方文学中的神奇性
Essay on the Marvelous in Western Literature
1927

这段写于20世纪初期、截取于米歇尔·莱里斯评论中的文字，说明了人类对"神奇事物"的需求，甚至成为一种必要性。它同时也指出"奇异性"出现的各种背景，犹如向平凡、沉闷、既定秩序而展开的挑战。

Written at the beginning of the 20th century to explain the need, even the necessity for the "Marvelous", this citation from an essay by Michel Leiris also defines the circumstances of the birth of a singularity as a way of fighting against banality, boredom, and the established order.

奇异性与美感一样，是一种非常个人化、难以定义和无法触知的概念。艺术家们百般追求，或者以生命直接经验着这种奇异性，以表达出其创作与想象力的独特之处。这个概念凭借着日常之中对新颖罕见、陌生好奇甚至离奇古怪的领域的探索，来自我滋养茁壮。

大自然也不例外，不论是植物或矿物，是否借助人为的介入，都孕育着自身的独特之处。

在广义的花园世界里，奇异性显然超越了多世纪以来人们为景观整治方式所界定出的古典美感与潮流。不论这些花园的创作者是造型艺术家、景观师、建筑师、音乐家、设计师……或者仅仅是"普通人"，他们作品的共通点在于对那些约定俗成的界限的突破。

Like beauty, singularity is a very personal, indefinite, impalpable notion. Artists woo it or simply live it in order to express the difference of their approach to creation, to imagination. This notion nourishes itself daily on the exploration of unknown, strange, unusual horizons.

Nature is not to be outdone, cultivating her own vegetable and mineral anomalies, with or without the help of humans.

In the domain of the garden, in the widest sense of the term, singularity obviously goes beyond the aesthetics and the classic tendencies of landscape design, such as they have been defined across the centuries. Garden creators are personalities in their own right. Whether multi-media artists, landscapers, architects, musicians, designers or simply the guy next door, they have in common the willingness to cross agreed-upon rules and ideas.

二平方米的永恒 /
2 m² of eternity - 2013
Laurence Garfield & Nathalie Houdebine
(法国)

Photos ©: Leila Garfield & Nathalie Houdebine

为了打破法国墓园的矿质特性，"二平方米的永恒"建议以传统的墓石以及能够象征性地唤起对死者记忆的花园，来取代一般的设计。此花园在灌溉上具有独立性，并且只需要极少量的维护工作，体现出生命的自然周期，不仅有利于创造城市的生物多样性，同时也将城市墓园转变为花园墓园……

Alternative to all the stone found in French cemeteries, "2 m² of eternity" proposes, instead of traditional tombstones, the creation of gardens symbolically evoking the memory of the deceased. Conceived to be low-water and low-maintenance, they evoke the cycle of life and participate in the biodiversity of the city and the metamorphosis of cemetery cities into cemetery gardens.

罗马尼亚山地聚落 / rou - 2010

Ciguë, Diplomates & Alexandre de Dardel (法国)

Images ©: Ciguë, Diplomates & Alexandre de Dardel

在罗马尼亚中心的布拉索夫，五栋住宅巧妙地利用一块15公顷基地的地形，形成一个完全融入野生山地景观的小聚落。这些住宅犹如托尔金小说《指环王》当中虚构的霍比特人房子，仿佛从丘陵中长出来一般，其边缘设置着铺板平台，将四周自然环境里的如画风景尽收眼底。

Inscribing themselves into a wild and mountainous landscape, five houses form a hamlet, aptly utilizing the topography of this 14-hectare site in Braşov, in the centre of Romania. Like the fictional houses of Tolkien's hobbits, they seem to emerge from the hills. At their edges are wood-planked terraces that greatly enlarge the opening onto the surrounding natural scenery.

他们转化、改变和超越人们对花园与自然的惯有概念，展现创意独到的眼光，有时令人疑惑，却经常是乌托邦式的，甚至包罗万象的。他们以不同规模的空间将参观者、过路人、街坊邻居们……带领到充满诗意、趣味与魔力的新奇世界，时而弥漫着质询精神、积极性与互动性。

这些奇异花园散发着自由精神，强调距离效果，重视差异性、环境、欢愉……它们赋予空间集体性格或私密氛围，并根据各自的特性而启发人们的深思与好奇心。

对花园的设计者而言，这些超乎寻常的避风港或独特象征提供了人们分享与交流的场所，也像是一场逃逸与梦想的邀约，为当今世界重新带来活力与魅力。

These singular gardeners divert, metamorphose, and transcend the notion of garden, and even of nature, offering their original vision, which is sometimes disconcerting, often utopian, even universal... They take the visitor, the passer-by, the neighbour to a disorienting elsewhere, where poetry play and magic meet up, but also where interaction, activism and questioning are encouraged, no matter what the scale.

Singular, or unique, these garden speak of liberty. Trying hard for detachment, they are concerned with otherness, with the environment, and with pleasure, but also with the collective as well as the intimate, arousing curiosity, both active and contemplative, by playing with their particularities.

These havens, these uncommon symbols, are places of sharing and exchange for their designers, each in their own way. Invitations to escape, to dream, they re-enchant a world that has great need for it today.

绿色岛屿 / green island - 2014
Parc de Jiuzhaigou (中国)
Photos ©: Sylvie Henrionnet

位于中国四川省的九寨沟国家公园被世界教科文组织列为世界遗产，同时也成为生物圈自然保护区。九寨沟里分布着九个藏族村寨，倘佯在108个湖泊之间。这些湖泊来自于遭雪崩阻断的河流，而青藏高原的石灰岩则使得湖水呈现出令人惊异的祖母绿颜色，并且极为晶莹清澈，让人可以看到水底及其非凡出众的海洋花园。漂浮于海洋花园之上的树枝群承载着一个个绿色岛屿。

Jiuzhaigou National Park, in the Sichuan region of China, has been classified as a UNESCO World Heritage City and Biosphere Reserve. Its name means "Nine Village Valley", for the nine Tibetan villages along its length, themselves surrounded by 108 lakes created when avalanches interrupted the flow of rivers. Thanks to the calcium carbonate found in the Tibetan Plateau, the water is an astonishing emerald colour and of a rare transparency, allowing one to see to the bottom of the lakes, with their amazing marine gardens, over which grow little green islands floating on branches.

虞美人 / poppies - 2010

Laurence Garfield (法国)
Photos ©: Leila Garfield

此花园具有瞬息即逝的特质，却又拥有无穷尽的更新能力，因此劳伦斯·加菲尔德从2010年起便让他的巨型虞美人四处流浪，从瑟里西国际文化中心到塞纳－马恩省的黄磨坊花园，再到巴黎第12区的空中绿色步道。虞美人的搪瓷花冠，既有火红艳丽的也有深暗婉约的，它们带着诗意与趣味到各处逗留一阵，仿佛随着风的动向自由流动。

Gardens must confront the ephemeral and contain an exhaustible power of renewal. Because they share these very qualities, the giant poppies of Laurence Garfield have wandered since 2010. From the International Cultural Centre of Cerisy to the Promenade Plantée that overlooks Paris's 12th Arrondissement, as well as the Garden of Moulin Jaune (Yellow Mill) in Seine-et-Marne, these flamboyant or understated poppies, of enameled terra cotta, stand poetic and playful, as if carried by the wind.

这个极具艺术性的方案为立体花坛的创作带来全新的面貌，在发型师达维达·恩希阿的参与下，几块花坛中的禾本植物像头发一样被编织、打结，成为具有原创性的黑人发型。从17世纪末到19世纪初，南特港曾经是奴隶贩卖中心之一，这个花园设计则借此机会向奴隶贸易条约受害人民的文化致意。

This project of artistic mosaiculture renews the genre. For example, grasses of several beds of the botanical garden are braided, knotted, becoming original afro hairdos. With the assistance of stylist Davida Nsiah, the plaited grasses render homage to the cultures of people victim to the black slave trade, of which the port of Nantes was one of the centres from the end of the 17th century to the beginning of the 19th century.

发辫艺术 / twist in cocody - 2010

Marie Denis (法国)
**Jardin des Plantes &
Musée des Beaux-arts de Nantes** (法国)

Photos ©: Marie Denis

01
植物之歌
VEGETATION

植物之歌
vegetation

迷宫 / labyrinth - 2008
France de Ranchin (法国)
Ministère de l'Éducation nationale (法国)
Photos ©: France de Ranchin

本项目位于上马恩省朗格勒高原的高洪森林中，弗朗斯·德·郎尚为了配合一项教学计划，在一片2000平方米土地上以立桩的方式呈现出她的迷宫。这些木桩围合成的死胡同和小径构成了迷宫的最初形式，之后黄杨树将取代它们，使迷宫融入到周边的景观中。

Within the framework of an educational project, France de Ranchin wrapped her design over 2 000 m² in the forest de Cohons, on the Langres plateau, in Haute-Marne. The plastic fencing offers a scenographed, original version of culs-de-sac and paths to explore, before the bushes encroach, melting the labyrinth into the landscape.

从古代起，自然与人工就在花园中形成既对立而又相辅相成的关系，植物则成为主要的象征性载体。几世纪以来，人们通过植物在景观空间中的形式、角色以及树种的选择，主导了当时的景观发展趋势，以美学的、哲学的、社会的、宗教的方式呈现出其时代和地域的特色。

植物与人类发展紧密相关，经常成为塑造场景的元素，不论是在以蔬果生产为主的花园里，或者在一些游赏性质较强的场所，它们的角色都介于自然与耕种文化之间。

20世纪初期的西方世界，花园的创作方式不再是单一的，不再只有一种主导风格成为独断的学派，而是多种风格平行地共同发展。历史上曾经出现的形式被重新加以审视，其中的一部分与新风格结合在一起。同时，在刚刚出现的抽象绘画的启发下，非常现代、简约而且几何化的花园形式出现了。

Since antiquity, nature and artifice compete and cooperate in gardens, vegetation being the principal symbolic medium. Through the centuries, the form of the vegetation, its place in the landscaped space and the choice of the species is going to determine the trend that will indicate its time in history, one connected in an aesthetic, philosophical, social and religious manner to the different epochs and relevant parts of the world.

Intimately associated with human development, vegetation is "scenographied" to participate, for some of its qualities, in the necessities of subsistence as crops, or to adorn the privileged leisure spaces between nature and culture.

At the beginning of the 20th century, in the Western world, garden design is not monolithic, there is no longer one dominant style with a big following, but many that develop parallely. Historic forms are revisited, for some in a new current. At the same time, very modern, sober and geometric forms are born, inspired by the pictorial abstraction that is making its appearance.

从1930年代起，景观设计从花园尺度急剧过渡到城市和大地景观的尺度，成为重要的社会议题，并且与大众息息相关，而不再只关乎精英阶层。人们开始谈论"绿化面积"和"绿色区域"，种植树木变成象征性的行动，到今天亦是如此

但是从1970年代开始，在生态意识抬头、生物多样性受重视、生活方式演变的这些背景下，人们见证了"花园的返回"。从此，景观设计开始变得多样化，是跨学科合作的构思者们想象力的结晶，植物也重新成为灵感的源泉。

在这些既注重形式又强调感性的景观实验所带来的丰富变革中，树木占有极为重要的分量。它们不再局限于线状种植，也由于其所具有的各种自然特性而受到重视、修整和欣赏。它们也被视为保护元素或者是某个观点、社会团体的整合者……

各类植物被应用在不同项目中，以唤醒人们的感官知觉，带来宁静与诗意的氛围，甚至崭新的互动，同时也以贴切而独特的形式，为人们创造惊喜或者传递关于生态的、社会的和其他方面的信息。

Starting in the 1930s, the idea of landscape, passing suddenly from the scale of the garden to that of the urban landscape, or of the territory, is going to become an important social concern, involving everyone, no longer just the elite. One talks about a "green surface" and a "green zone" where the planting of trees is emblematic, as it still is today.

But during the 1970s, some went "back to the land" with a foundation in ecological awareness, in the necessity of biodiversity and in the evolution of lifestyles, habits, desires. Since then, landscape design has been multiple, fruit of the imagination of multi-disciplinary artist for whom vegetation becomes once agains a source of inspiration.

In the midst of these movements rich with formal and sensory experimentation, the tree takes an important place. It is no longer relegated to alignments but honoured, adorned, admired for its natural particularities. It is also identified as a protective element or a unifier of an idea, of a community

The whole of the vegetable world is generally used, according to the project, a call to the senses, a serene, even poetic, ambiance or a innovating interactivity, but also to question, to surprise, to deliver an ecological or social message... and to re-enchant the world by clear and singular forms.

苏比亚科的椭圆形庭院 /
the subiaco oval courtyard - 2014
Luigi Rosselli, Kristina Sahlestrom (澳大利亚)
Photos ©: Edward Birch

这栋别墅位于珀斯西部郊区的苏比亚科街区，建造于20世纪初期。为了重新赋予利用价值，重修的所有建筑都围绕着一个椭圆形的庭院和游廊而组织。这个庭院花园被刻意以简约朴素的方式处理，偏离中心点的一棵参天大树扮演着统领者的角色，所有室内空间都能直接到达院子并且拥有直接的视野。

For the redevelopment of this house dating from the beginning of the 20th century, in Subiaco, a suburb west of Perth, everything was conceived around an oval courtyard, echoed by the surrounding elliptical veranda. This intentionally sober garden functions like a magnet, all the rooms having both physical and visual access to it.

听…… / listen... - 2003

Olga Ziemska (美国)
Centre of Polish Sculpture (波兰)
Photos ©: Olga Ziemska

这件位于波兰欧隆思科的波兰雕塑中心的作品，展示出艺术家奥尔加·杰姆斯卡与大自然之间的特殊关系。石膏制成的手形雕塑被"移植"在瘦长的桦树干顶端，此般奇特的造型，堪称为超现实主义作品。但它实际上是以一种极为个人化的形式语言在传递信息，提醒人们如果没有周围的世界，我们将一无是处。

This work, at the Centre of Polish Sculpture in Orońsko, in Poland, illustrates the particular relationship that the artist, Olga Ziemska, has with nature. One could label it surrealist for its strange form, hands of plaster grafted to thin birch trunks. But it is in fact a message, transmitted by a language of very personal signs that recall that we are nothing without the world that surrounds us.

韦尔奇山毛榉
the faux of verzy
法国 France

长久以来，一千多棵变形的山毛榉就生长于兰斯山脉的森林中，形成一项奇特的自然遗产。它们以相邻一个村庄的名字命名，这些自由生长的山毛榉犹如活生生的雕塑，点缀着这片广阔的森林。

弯曲的枝干互相连接在一起，大量的吻合现象令人迷惑，成为科学家的不解之谜，同时又引发无数的想象和猜测：中世纪僧侣实施的无性繁殖、神的惩罚、土壤里的放射元素、地下的通风气流、大地辐射……总之，这是一片魔幻之地。

A strange natural patrimony has forever populated the forest of the Mountain of Reims. More than a thousand twisted beeches, called the "Faux of Verzy", from the name of the neighboring village, grow there spontaneously, punctuating this big sylvan space with strange living sculptures.

The importance of this phenomenon of anastomosis, with the branches bending and braiding together, fascinates us yet remains an enigma for the scientists, thus giving birth to many legends and speculations: monks planting them during the Middle Ages, divine punishment, radioactivity of the soil, subterranean air currents, telluric radiation… A site where magic reigns.

Office National des Forêts
法国
Parc Naturel Régional de la Montagne de Reims
法国

Photos ©: Sophie Barbaux

森林小径
a path in the forest
爱沙尼亚 Estonia

2011

这条优雅的天桥飘浮在爱沙尼亚的卡林柯治森林中，围绕着那些百岁树木重形成一条新颖而特殊的散步道。与枝干和树叶的更近距离接触改变了人们与大自然交流所产生的认知和感觉，并随着季节发生变化。

This elegant footbridge, floating in the forest of Kadriorg in Estonia, reinvents the promenade in the heart of the more than hundred-year-old trees. The proximity of the boughs and the foliage modifies one's perception and the feeling of communion with nature, changing each seasons.

Tetsuo Kondo Architects
日本

Photos ©: Tetsuo Kondo Architects

身体景观
the body landscape
法国 France

2009

这些照片展现出法国利摩日地区一个青少年工作室的成果，以独特的方式讲述了一次森林模拟的体验。这个游戏叫做"我觉得我是……一棵树或者一丛灌木"，游戏规则要求参与者现场采集植物并覆盖在自己的身体上。当然，得屏神静气而又骄傲地展示自己，仿佛自己真的是那棵树或者那丛灌木一样，这些模仿萨满信仰的童稚装扮为孩子们带来一段共享的美好时光！

Fruit of a workshop of young children of the Limousin region, these photographs recount in their fashion an experience of woodland empathy. The rule of the day is to gather greenery and cover yourself with it, to play make-believe "that you are... a tree or a shrub". And certainly, to be quiet and stand proudly, as trees do, for a shared moment of child-like shamanism.

Marie Denis
法国
CIAP de Vassivières
法国

Photos ©: Marie Denis

过道
passage
德国 Germany

2007

这个用树枝堆砌起来的门洞犹如悬浮在林中空地的对面,越靠近反而越不清楚它是否在向游人发出邀请,因为并没有从中穿过的小路,一片茂密的森林封闭了门之后的空间。所以这个过道更像是一种违反常规的展示或者是一项仪式,邀请人们对于自己"处身大自然中的方式"提出质疑。

The closer one comes to this threshold of branches and twigs, a sort of levitating entrance, the invitation to cross over becomes less clear: no path and a dense forest close the space beyond. One is lead rather to a transgression, to a ritual that questions our place in the heart of nature.

Cornelia Konrads
德国
Sculpture Landscape Osnabrück / Association for New Art TOP.OS
德国

Photos ©: Cornelia Konrads

惊奇柴草堆
impressive haystacks
法国 France

2013

此方案旨在向一些已经从今日景观中消失的自然建构物致敬。这些以传统方法搭建的柴草堆矗立于花园中，不仅见证着一个没有人为修饰的旧世界，也向参观者展示它们雕塑般的大方形态，让人们回到过去时光中散步和休息。

Homage to those natural architectural structures disappeared from our landscape, these haystacks, traditionally made, rise up, testifying to a bygone world without artifice. They offer their generous and sculptural forms to visitors, who immerse themselves in this bygone world, the time for a promenade, for a break.

Robin Godde
法国
Festival International des Jardins Domaine de Chaumont-sur-Loire
法国

Photos ©: Robin Godde p.26
Sophie Barbaux p.27

"球茎"沃土
fertile bulbs
法国 France

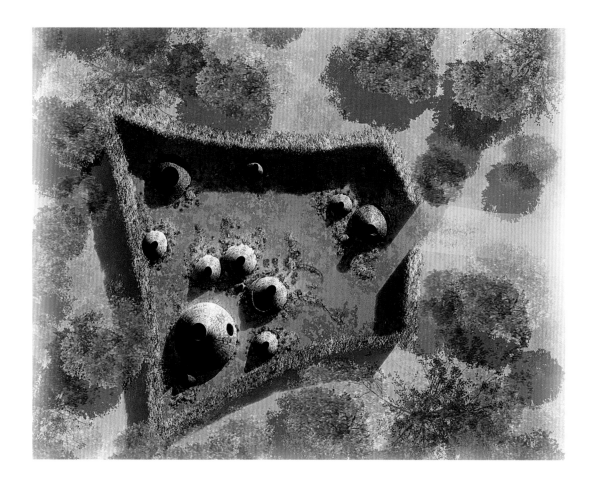

"明日花园与巧妙的生物多样性艺术"是2011年卢瓦尔河畔－修蒙国际花园展的设计主题，本方案特别推崇土壤的饶沃与自发力量，展现腐殖土以其分解周期来造化生物的自然神奇。

一些由杨木条编织而成、形体丰满圆鼓的巨型"球茎"像是被人随意扔在花园尽头，就此落地生根，并带来出人意表的成果。

"Future gardens or the art of happy biodiversity" is the theme, that year, of the International Garden Festival of Chaumont-sur-Loire. In this landscape project, the richness and the spontaneous work of the soil, the natural magic of the cycle of humus, have the place of honour.

As if flung to the back of a garden, surprising giant bulbs grow, with rounded forms, braided with poplar strips normally used for the making lightweight crates.

Mésostudio / Xavier Bonnaud & Stéphane Berthier, Fabien Gantois Architectures, Les 2 Cyclopes, L'envers du jardin
法国
Festival International des Jardins Domaine de Chaumont-sur-Loire
法国

Photos ©: Mésostudio / Xavier Bonnaud & Stéphane Berthier, Fabien Gantois Architectures, Les 2 Cyclopes, L'envers du jardin

"球茎"内装了半满的肥沃基土,提供种子自发萌芽的环境,一如任何堆肥中的种子,在吸收有机物质之后便顾自发芽茁壮。旱金莲、翼叶山牵牛、芳香植物、葫芦、面条瓜等,这些不论是攀爬、上窜或蔓延的植物,不仅盘踞了"球茎"表面,也逐渐侵入四周空间,仿佛演奏着一曲对植物繁殖力与生物多样性的颂歌。

Half full of a rich substrata, they are the theatre of a spontaneous germination, as happens in all compost where seeds, fed with organic material, germinate like crazy. Climbing, spurting or running, nasturtiums, black-eyed Susans, aromatic herbs, spaghetti squash and gourds invade the surrounding space progressively, creating an ode to domestic and underground fertility and diversity.

植物装置
stick work
2003-2012
澳大利亚, 意大利, 美国 Australia, Italy, USA

这位精通编织和织造艺术的大地艺术景观师把他的巨型构造物散布在大自然之中，而每一次大自然又成为其创作的源泉。艺术家受到鸟类筑巢原理的启发，采集当地生长的柳树、榛树、槭树、榆树或者其他年轻树种的柔软枝条，并把它们像鸟巢一样交错地编织成蜂巢或者蚕茧、原始的茅屋、巢穴，甚至是宫殿，所有为想象力而开放的庇护所。

Master of the arts of braiding and weaving, this sculptor of land art, installs his gigantic architectures in a nature continually explored as the source of creation. Inspired by birds' nest-building, Dougherty bends supple branches of willow, hazel, maple, elm or other saplings taken from the area, and weaves them together to make beehives, cocoons, simple huts, dens or even primitive palaces.

Patrick Dougherty
美国

野生庇护所 /
Na Hale 'Eo Waiawi
Contemporary Art Museum
美国檀香山, 2003
Photos ©: Paul Kodama

无论它们看起来是多么美若仙境或者异想天开甚至令人不安，这些巨大的植物雕塑都如同真实的梦境一般存在于环境中。设计者帕拉提克·多尔蒂在北卡罗琳娜森林中度过的童年让他与树木产生了默契。在每项作品的实现过程中，他都邀请志愿者一同来分享建造过程以及这份与大自然的默契，他们一同采集、切割枝条并把它们分类，之后再编织成那些昙花一现般的作品。

Whether they are fairy-like, whimsical, even disquieting, these immense vegetal sculptures integrate into their environment like dreamed reality. For each realization, Patrick Dougherty invites volunteers to share his construction work and his collaboration with trees, heritage of a childhood passed in the forests of North Carolina. Together the volunteers and he gather, harvest, cut, assemble and tie the boughs, before erecting them and giving form to ephemeral works.

避暑宫 / Summer Palace
**Morris Arboretum
University of Pennsylvania**
美国费城, 2009
Photos ©: Rob Cardillo

小型宴会厅 / **Little Ballroom**
Federation Square
澳大利亚墨尔本, 2012
Photos ©: Megan Cullen
p.36

就在街角 / **Just Around the Corner**
New Harmony Gallery
美国新哈莫尼, 2003
Photos ©: Dole Dean
pp.37-38

共享家园 /
Tana Libera Tutti - Home Free
Arte Sella
意大利瓦尔苏加纳, 2011
Photos ©: Giacomo Bianchi

普克树
pooktre
澳大利亚 Australia

1987

珠宝设计师皮特·库克在昆士兰州东南拥有一座65公顷的果园。1987年，他萌发了一个念头：把柳树插条栽培成一个椅子的形状。接着，在工艺美术家贝姬·诺西的协助下，这个艺术探险于10年之后在他自家果园中获得了实现。

他们共同合作，发展出一种令人称奇的树木整形方法，尤其适用于某一类变种的野生李树，一种被称为樱桃李的当地特有树种。他将这个方法起名为Pooktre（普克树），是皮特·库克的小名Pook以及tree（英语"树"的意思）这两个词的结合。

Everything began in 1987 for Peter Cook, a jewelry designer with the idea of "growing" a chair by planting willow cuttings... Ten years later, with the collaboration of the stylist Becky Northey, this simple act becomes an artistic adventure in the orchard of his property of 65 acres, in the mountains of southeast Queensland.

Together, they have developed an astonishing method of sculpting the trees, more particularly using a variety of wild prune trees endemic to their region, the *Cerasifera myrobalan*. They call it Pooktre, combining Peter Cook's nickname, pook and tree, into a single word.

Peter Cook & Becky Northey
澳大利亚
Queensland
澳大利亚

Photos ©: Pooktre

此方案的原创性在于他们并不只是局限于改造已经具有一定外观基础的树木，而是以确切的造型为目标而创作，甚至让这些树在未来具备一定的功能：桌子、镜子、椅子……甚至让人浮想联翩的人物造型都在设计范围内。这些树木伴随着季节更迭而产生变化，春季繁花似锦，夏季果实累累，秋季金黄缤纷，在冬季则与形态简约的阴影相映成趣……

The truly original aspect of their work is that they do not construct using existing material. They develop a precise figurative, even functional, project by planting and giving form to materials that are in the process of becoming. Tables, mirrors, seats are born in this way, as are characters with evocative silhouettes that metamorphose each season, enchanting with the first buds of springtime, laden with fruit in the summer, glowing with autumnal colours, then playing with their shadows, pared down in winter...

树木房客
tree tenants
奥地利 Austria

1973

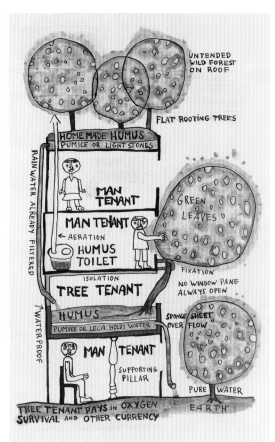
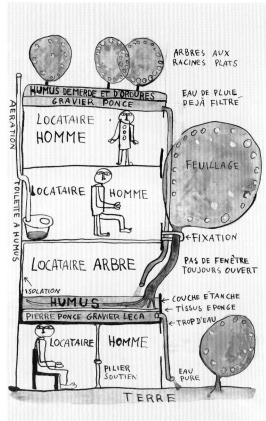

除了画家、思想家、建筑师这些头衔之外,百水先生早在1990年1月24日的宣言中便把自己称为"建筑医生",他从很早之前就开始在绘画作品、出版物以及建筑设计中表达他的生态理念。他相信乌托邦的现实性,并把方案付诸实践,这个从1980年开始发展和实施的前卫项目"树木房客"就是个很好的例证。此方案重新赋予那些"植物大人"们最适当的位置,成为人类不可或缺的自然伙伴。

Painter, thinker and architect, or more precisely, "doctor of architecture" as he calls himself in his manifesto of January 24 1990, Hundertwasser expresses very early his ecologist ideas in his pictural, written and architectural work. He puts his utopias, that he sees as living proof, into practice, such as in this avant-garde work, Tree Tenants, started in 1980. It restores the lord of vegetation to its rightful place, natural and indispensable partner of Man.

Friedensreich Hundertwasser
奥地利

树木房客的766张草图, 1976
766 Tree-tenant drawings, 1976

Photos ©: 2014 Hundertwasser Archive, Vienna

**百水先生之家 /
Hundertwasser House
Residential building of
the City of Vienna
Original coauthor : Josef Krawina**
(em. o. Univ.-Prof. Arch. DI)
Planning: Peter Pelikan (architect)
奥地利维也纳,1983-1985
左/left

**维也纳美术馆 /
KunsHausWien**
奥地利维也纳,1991
右/right

这些树木被种植在专门为此预留的空间中，距离住宅窗户一米远的地方。百水先生反对城市污染和森林砍伐，主张把原本属于自然的空间归还给自然，他同时也反对传统的灰暗而又阴沉的建筑立面。这些城市中的森林代言人比人类付出更昂贵的租金：它们创造和释放氧气、改善周围空气的质量和舒适度、吸收各种有毒元素、减少噪声并且遮挡外部视线，这一切都使得它们的同居者受到保护并且拥有宁静的生活环境。

By planting these trees a meter from the windows of the residences, in spaces conceived specifically for this end, Hundertwasser fights against urban pollution and deforestation, restoring to nature its place, but he also combats the classic drabness of the facades. These forest ambassadors to the city pay higher rent than men by creating oxygen, and improving the ambient air and the well-being of the inhabitants. They absorb various toxic particles, reduce noise and also protect homes from exterior view, giving shelter and calm to the human tenants.

红地毯！
red carpet!
法国 France

2011

这个实地装置的作品是为了庆祝"艺术与自然之路，我和一位艺术家在小径上相遇……"项目的十周年纪念。象征至高敬意的"红地毯"蜿蜒穿过若亚克小镇，连通不同区域的自然空间，也把市镇中心和山谷连接在一起。这条绿色草地如同一丝从沥青路面中迸发出的自然活力，伴随着人们日常中的漫步。

An installation in situ to celebrate ten years of the project "Path of Art and Nature Path, along my route, an artist passes..." Red carpet!, veritable symbol of homage, unfolds through the village of Jaujac inking natural spaces on one side and the other, connecting the heart of the town with the valley. Snaking from streets to paths, stairs to alleys, this lawn accompanies daily amblings, like a wisp of nature, resurgence of the living on asphalt.

Gaëlle Villedary
法国
Sentier Art et Nature de Jaujac
法国

Photos ©: David Monjou

猫头鹰
owl
USA

Edina Tokodi / Mosstika
美国

Photos ©: Mosstika

莫斯迪卡街头艺术家团体运用绿色草皮在城市中重新描画出动物寓言,以他们独特的方式来对抗城市空间的冷漠与枯燥无味。这是送给纽约市民和游客的礼物,同时也使得这个大都会诞生了一个受人欢迎的"绿色游击队"。

Fighting in its own way against the hardness and sterility of urban spaces, Mosstika reintroduces to the city a bestiary of two-dimensional grass animals affixed to walls and fences, offered to passers-by and New York residents. Much welcomed, a green graffiti, is born, on scale with this megapolis.

巨人头与泥土少女
the giant's head and the mud maid
1998
英国 United Kingdom

这座位于康沃尔郡的公园是最富盛名的英式植物园之一，从16世纪创建之后，陆续增加的一系列游赏花园、蔬果园以及保护森林使其逐渐丰富起来。这些令人称奇的巨大黏土人物栖身于一片绿意中，好像从泥土中直接生长出来一般，带给游客无比惊喜。

Situated in Cornwall, this park, one of the most famous British botanical gardens, was created in the 16th century, then enriched over time by different pleasure gardens, vegetable gardens and protective woods. Nestled in a framework of green, fantastical characters modeled in clay surprise visitors by seeming to emerge from the caverns of the earth.

Sue et Pete Hill
英国
Les Jardins perdus de Heligan
英国

Photos ©: Heligan Gardens Ltd

图尔坎墓园
the cemetery of tulcan
厄瓜多尔 *Ecuador*

墓园里的上百棵柏树种植于1936年，在管理主任何塞·马里亚·阿扎里·佛朗哥的精心照料下，这座墓园已成为世界上最大的修剪艺术花园之一。经过定期修剪、平整和雕饰，这些柏树变换成300个造型各异的人物、前哥伦布时期的神灵，甚至令人称奇的动物……目前，墓园主任的儿子继承了父亲的工作，继续这项具有创意的园艺探奇。

Planted in 1936 by José María Azaèl Franco, director of the cemetery, more than a hundred carefully sculpted cypress trees form what is probably the biggest topiary garden in the world. Regularly cut, shaved and combed are 300 astonishing human or pre-Columbian figures, divinities and fantastic animals. Today, Franco's son continues this creative gardening adventure.

José María Azaèl Franco
厄瓜多尔

Photos ©: John Meckley

植物之歌 / 雕琢

拉巴鲁城堡花园
the gardens of the château de la ballue
法国 France

1973

这座布列塔尼边境省的旧堡垒位于距离圣米歇尔山只有几链（旧时计量距离单位，一链大约200米）的地方，在17世纪被改造为城堡。从1973年开始，出版人克劳德·阿尔托在乌托邦主义建筑师弗朗索瓦·赫伯特-史蒂文斯和保罗·梅蒙德的协助下，赋予城堡里的法国式园林一个崭新的面貌，这两位建筑师分别设计了17世纪意大利古典花园以及呈现出众多惊喜的风格主义的对角线花园。克劳德·阿尔托甚至在勒·柯布西耶某个项目的启发下，重新种植了1 500棵紫杉树，并使它们围合成迷宫。

A stone's throw from Mont Saint-Michel, the French gardens of this 17th century château built on an ancient fortress of the Breton Marches, has been given, beginning in 1973, a breath of fresh air by the editor Claude Arthaud. Helping her in this project were Utopian architects, François Hébert-Stevens and Paul Maymont, who each designed a garden space, one in the style of a 17th century Italian garden and the other a Mannerist diagonal garden of thirteen surprises. Ms. Arthaud notably replanted a labyrinth of 1 500 yew trees, inspired by a project of Le Corbusier.

Marie-Françoise Mathiot-Mathon
法国

Photos ©: Yann Monel

这些奇妙而又精致的花园以深浅不一的绿色调、光影的对比，以及周围博卡日田园中升腾起的薄雾来强化景观效果。从2005年开始，它的新主人玛丽-弗朗索瓦兹·马蒂奥-马东在这片烙印着诗歌和爱丽丝仙境氛围的土地上继续进行花园创作。她在当中设置了马蒂纳·萨拉维萨和索菲亚等艺术家的雕塑作品，致力于保持花园的独特之处，同时也通过摄影师晏·莫纳尔的影像作品为城堡花园出版图书，为这个永远超越时间的场所留下纪念。

These unexpected and refined gardens play with plaids of green, contrasts of shadows and light, as well as with the mist rising from the surrounding hedgerows. Since 2005, its new owner, Marie-Françoise Mathiot-Mathon, has continued work on this site so imprinted with poetry and with an Alice in Wonderland atmosphere. Here, she gives space to the sculptures of artists such as Martine Salavize and Sophia, and has enhanced the uniqueness of this garden by publishing a book about it, aided by the photographer Yann Monel. Forever outside of time.

56 vegetation / SCULPTED

痕迹
tracks
意大利 Italy

2001

格拉保别墅位于圣庞加爵的卢卡城边,这件装置作品就在这座别墅的历史公园中展开,层层叠叠的翠绿色波浪向园林修剪艺术和立体花坛致敬,它意味着人类对手工技能的喜爱更甚于自然本色,并以一种具有绝然当代感的形式重新诠释祖传的技艺。

This installation deploys its multiples emerald waves in the historic garden of the Villa Grabau, in San Pancrazio, near Lucca. A homage to topiary art and to mosaïculture, the garden symbolises human preference for artifice over the natural and revisits under a resolutely contemporary form, its ancestral savoir-faire.

Roberto Capecci, Marco Antonini, Raffaela Sini, Land - l archicoltura
意大利
Festival di Arte Topiaria, Lucca
意大利

Photos ©: Land - l archicoltura

58 vegetation / SCULPTED

绿色铸造
green cast
日本 Japan

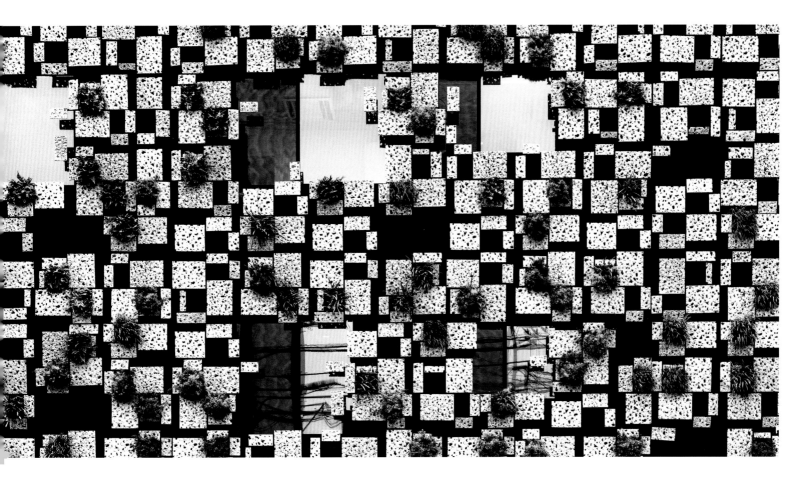

Kengo Kuma & Associates
日本

Photos ©: Kengo Kuma & Associates

这座多功能建筑位于日本小田原市,其广阔立面由错落有致的压铸铝板拼接而成,呈现出不规则的样貌,看起来更像是石材所造。栖身其中的植物点缀着这片巨大而又零散的方格网,这个垂直花园继承了20世纪初期的前卫风格,显得既严谨又有机。

The facade of this multi-purpose building, in Odawara-shi, in Japan, is made of an immense patchwork of molded aluminium panels, with irregular, almost mineral, features. The plants nesting there punctuate this large, jumbled chessboard, with shapes inherited from the avant-garde of the beginning of the 20th century, offering a vertical garden, simultaneously rigorous and organic.

60 vegetation / SCULPTED

摩根城之树
the morgantown tree
美国 USA

2013

作为"毛线涂鸦"的先驱者之一,卡罗尔·于梅尔为树干和树冠都穿戴上紧身毛衣,以此向这些树木致意。树木是生命的象征,同时也让我们思考自己与这个世界、与其他人以及与自然之间的关联。卡罗尔·于梅尔的每一件作品都跨越了文化的界限,成为建立社会联系的元素,它们迷人、令人称奇甚至饶有趣味,吸引着那些不同年龄、性别、宗教信仰、政治派别、种族和教育程度的志愿者前来共同进行创作。共有100个人参与了摩根城之树的编织,他们用了一个月的时间,一起创造了西弗吉尼亚州社会团结的象征。

Among the pioneers of "Yarn bombing", Carol Hummel covers the crowns and trunks of trees with tight-fitting and crocheted clothes: homage rendered to the trees, symbols of life, making us question our relation to the world, to others and to nature. Crossing cultural frontiers, each of her works aims to be participatory, part of social ties, by engaging, surprising, even amusing the associated volunteers, of all ages, sexes, religions, political affiliation, race and education. Thus, 100 people participated in the realisation of the month-long The Morgantown Tree, together creating an emblem of the unity of their West Virginia community.

Carol Hummel
The Morgantown Tree
West Virginia University
美国

Photos ©: Carol Hummel
p.62 & p.63 上/top
Daniela Londoño
p.63 下/bottom

疯狂之草
wild grasses
法国 France

2010

在一片茂密森林的中心地带，这些尺度巨大的草似乎来自于格里弗游记中的巨人国布罗丁那格，伸展着独特的茎干。它们的表面装饰着成百上千块马赛克碎片，通过不同的构图、材质和颜色向人们讲述微观动物世界的多样性、不稳定性和拟态性，那些动物通常隐藏在枝叶间以寻求保护。

At the heart of a dense forest, high grasses prick up their singular stalks, on the scale of Brobdingnag, the city of the giants in Gulliver's Travels. Adorned with thousands of mosaic tesserae, these grasses narrate through their compositions, their textures and their colours, a story of diversity and fragility. They highlight the mimetism of the miniscule animal world, usually sheltered, hidden among the blades.

Alexandra Carron
法国

Photos ©: Alexandra Carron

索朗日
solange
法国 France

Claude Cormier & Associés
加拿大
Domaine Lacroix-Laval
法国

Photos ©: Annie Ypperciel

Image ©: CCAPI

在里昂附近的拉克鲁瓦－拉瓦勒国家公园中，这片树林被装饰得像是要去参加舞会，条带状的裙子和人造的丝绸花朵吸引着游人的注意。在细长树干的强调下，这个垂直的美景使整片森林的层次更加清晰，并提醒人们这个地区关于丝绸纺织的重要历史。

Dressed as if going to a ball, a grove of trees at the Park of Domaine Lacroix-Laval, near Lyon, draws attention to its gowns of ribbon and artificial silk flowers. This vertical fairyland showcases the structure of the forest space, rhythmed by the linearity of the trunks, all while recalling the historical importance of the spinning mills of the regional silk industry.

植物的婚礼
vegetal wedding
法国 France

巴黎植物园的大温室在一段展览的期间里，化身为一个浪漫剧场，带人进入一场认识婚礼仪式与象征的启蒙之旅。设计师以令人惊叹的拟态手法，用取自硅元素的天然材料制作的白色花边，来呈现未来新娘的各种表征物件，这些装置使得这个受到保护和精心管理的自然景观得到升华。

The setting of the Grandes Serres du Jardin des Plantes (the Large Greenhouses in the Garden of Plants) in Paris, has become, during the length of an exposition, the stage for a initiatory voyage through the rites and symbols of marriage. The traditional attire of the bride is imitated with disturbing realism by white lace created through a natural material made of silicium, thus enhancing a protected and managed nature.

Tzuri Gueta
法国
Fondation Ateliers d'Art de France & Muséum national d'Histoire naturelle
法国

Photos ©: François Junot

游旋塔玛
the touring tama
爱尔兰 *Ireland*

2013

Miriam Mc Connon Papageorgiou
爱尔兰
Dublin City Council & Dublin City Parks
爱尔兰
Photos ©:
Miriam Mc Connon Papageorgiou

在塞浦路斯，人们会把一块手帕系在树上以示还愿，如今这项传统已经消失（在希腊文里这块手帕被称为塔玛）。为了纪念爱尔兰成为欧盟轮值主席国，从两个国家收集的手帕拼接成的这块巨大塔玛，被罩在都柏林梅林广场上一棵老梧桐树的树冠上，邀请人们向这棵奇特的许愿树献上编织的供品。

In Cypress, the tradition endures where one hangs handkerchiefs on a tree as ex-votos (tama in Greek). To commemorate the presidency of Ireland at the head of the European Union, a giant Tama, made of handkerchiefs collected in the two countries, has enveloped an old plane-tree in Dublin's Merrion Square, homage to the offerings woven on that strange wishing tree.

空中花园
aerial garden
法国 France

2009

此花园诞生于在巴黎杜伊勒里公园举办的"花园，花园"活动中，是罗兰百悦盛世香槟的写照。整个花园优雅流畅，却又井然有序、充满活力，作为标志的一棵大树上缀满着金色珍珠的叶子，让人联想起那些极富盛名的香槟气泡。地面覆盖着丝般柔顺的禾本植物，犹如朵朵云彩点缀着精致的花朵，在空气和光线的变化中时而轻微摇摆，时而唏嗦作响。

Conceived for the exhibition "Jardins, Jardin", on the terrace of the Tuileries in Paris, this garden represents a prestigious vintage, the "Grand Siècle" of the Laurent-Perrier champagnes. It is fluid and elegant, all the while structured and powerful, its tree emblematic, with multiple leaves in golden mother-of-pearl evoking the famous bubbles of champagne. Clouds of silk grasses envelope the ground below, scattered with delicate flowers, the ensemble vibrant, clinking with the slightest change in the air or the light.

Cao | Perrot
法国&美国
Champagnes Laurent-Perrier
法国

Photos ©: Stephen Jerrome

苍鹭之树
the heron tree
法国 France

2002-2019

若干年以来，一条庞大且绿意盎然的奇异枝干一直盘旋在南特岛机械城的屋顶之上。这棵直径50米、高30米、拥有22条枝干的巨树，是"城市之树"项目的原型和实验品，身上装置着一系列令人称奇的空中花园和露台。

A strange branch, giant and lush branch has perched on the ships of the Machines de l'île de Nantes for several years now. A prototype, it experiments with and also prefigures the project of a tree-city in the sky. 50 meters in diameter and 30 metres high, replete with 22 branches, the Heron Tree features a succession of incredible gardens and suspended terraces.

Les Machines de l'île
法国
Ile de Nantes
法国

Photos ©: Pierre Orefice p.70
Sophie Barbaux pp.71-73

巨树上的"苍鹭之旅"是可以活动的机械装置,带给人们一段真实的飞行体验,让人们联想起这种沿着卢瓦尔河筑巢的涉禽,特别是在邻近的苍鹭岛屿上。整个树冠上充满了各种惊喜,人们在此遇到一群散步中的奇怪动物"旅行角蝉"、一条把游客从一条枝干运送到另一条枝干的"尺蠖蛾毛虫",或者一群互相紧挨、沿着树干爬行的"巨型蚂蚁"。

Above, the circuit of herons, very life-like contraptions on which visitors can take a flight, evoke that wading bird inhabiting the edges of the Loire, notable on the neighbouring, eponymous island. The ensemble of tree-tops holds many more surprises, such as strange animals promenading: the Membracides de Voyage, or traveling tree-hoppers, a creeping caterpillar that guides the visitor from one branch to another, and a colony of giant ants, busy climbing all along the trunk, in Indian file.

儒勒·凡尔纳的发明、莱奥纳多·达芬奇的机器世界和南特的工业历史在这里交织在一起，这棵安置在老造船厂基地上的苍鹭之树正准备迎接未来……

A cross between the invented worlds of Jules Verne, the mechanical universe of Leonardo da Vinci and the industrial history of Nantes, the Heron Tree is built on the site of former shipyards. To be continued...

植物之歌 / 梦想

这些具有雕塑感与趣味感的喷水壶是为了向园丁致意而产生的作品,它们在日间是白色的,夜幕降临之后则变成灯笼,散发出珍珠般的光晕。这些喷水壶象征着那些为了维护我们日常生活空间和城市空间而持续不断进行的必要工作,同时也呼应着大自然的节奏。

These sculptural and playful watering cans render homage to gardeners. White in daytime and at night converted into lanterns from which escape pearls of light, the watering cans, in kinship with the rhythm of nature, symbolise the unceasing daily work necessary to preserving our gardens and urban green spaces.

守护天使 / guardian angels - 2012

Maro Avrabou & Dimitri Xenakis (希腊&法国)
Le parc du Futuroscope de Poitiers (法国)

Photos ©: Dimitri Xenakis

02

协奏三曲

ACCOMPLICES

协奏三曲
accomplices

调色板 / palette - 2014
Paul Cocksedge Studio (英国)
London Design Festival /
Great Festival of Creativity (土耳其)
Photos ©: Mark Cocksedge

受到伦敦设计节的邀请，这件装置作品是为了2014年在伊斯坦布尔举办的第一届创意英伦盛典的开幕式而创作的，用以庆祝英国与土耳其在历史、经济及文化之间的联系。设计师保罗·考克斯基把两国国旗的颜色融合在一起，创造出一种全新的颜色，作为世界经济增长背景下的设计重要性的象征。

Commissioned by the London Design Festival for the inauguration of the Great Festival of Creativity, the first of which took place in Istanbul in 2014, this installation celebrates the historical, commercial and cultural relationship between United Kingdom and Turkey. Melding the colours of their respective flags, Paul Cocksedge creates a new, hitherto unseen tint, symbol of the importance of design in global economic growth.

水以各种形式出现在花园中，不仅使人着迷、引人注目、凝聚人心，也对人们的各种感官产生作用。这说明了为何我们对水具有本能的和有选择性的好感，尤其在花园中。水是天空和大地的天然镜面，以映射的多重影像，强化了环境中的景观，不仅创造出幻影，同时也为人们带来轻松、愉悦和闲散的感觉。水在经过规划后可以涓涓细淌、淙淙奔窜、潺潺流滚……或者泉涌而出！它提供人们一个特殊而又丰富的声音世界，甚至在所有尺度上改变了人们对景观的感知。

这个作为生命象征元素之一的水，在今天成为重要的生态议题，不论设计师们是重新利用古老的系统或者发明崭新的方法来节约用水和循环再利用。某些设计师试着提醒人们关注水资源一些新的迫切性，倡导尊重这个需要被保护的自然元素。另外一些设计者则继续利用水来创造游戏趣味，同时强调水的永恒循环，并且将它向来带给人们的惊奇体验转化为崭新的创作诠释，以回应人们的集体记忆。

Water fascinates, attracts, unites under all its forms and acts upon our senses, which explains the instinctive and specific affinity that one has for it, especially in gardens. Natural mirror of the sky and the earth, water multiplies the surrounding landscape. It is a creator of illusion but also of relaxation, pleasure and escape. As a performer, water can be dripping, gurgling, murmuring, gushing! It creates a particular sonorous universe that enriches or rather modifies the perception that one has of landscapes designs, on all scales.

Also symbol of life, water has become today an important ecological issue that designers integrate, reutilising ancestral systems or inventing methods that convert or recycle it. Creators are now aware of the new necessities, campaigning for the respect of this natural element that needs to be protected. Others continue to play with water, while recalling its perpetual cycle and revisiting the wonder that it has always elicited, echoes of our collective memory.

光线也是自然地伴随着花园的重要元素，从黎明到黄昏，无论是日正当空还是贴近地面，它从冰冷的白色过渡到温暖的色调，随着季节的更替而发生转变，并且大幅度地改变了周围的环境。提到光线也就涉及阴影，或深暗或着色，或伸张或蜷缩，为人们演奏出一曲愉悦视觉的交响乐。

作为照明用途的光，长久以来就以火、蜡烛、煤气灯的形式存在，煤气灯逐渐被电灯和如今的太阳能灯取代。它为花园带来魔幻般的夜晚，呈现与白天截然不同的画面，形影婆娑之中，植物、步道与建筑都开始有了远近变化。光不仅创造美景，也可以成为信息的来源，用画面讲故事，或者产生一些新的用途。它通过不断更新的技术以满足越来越严苛的节约能源的环保目标。

风则是花园乐章的第三个协奏者，占有特殊而又矛盾的地位。它犹如一位无法看见也无法控制的敌人，总是促成保护结构的产生，以减免由它所引起的水应力或物理应力所带来的损失，它的强大力道也让土壤变得干燥，并且使植物弯腰。但是，我们也不能忘记风在花粉传播以及促进光合作用方面的贡献。它让花园充满动感，甚至歌唱起来，如同我们在某些花园采用的新形式装置中所看到的，风展现出正面的、创新的、诗意的一面。

Light accompanies the garden naturally from dawn until dusk, zenithal or blazing, going from a white coldness to a colour heat, transforming across the seasons and metamorphosing its environment. Whoever says light says shadow, dark or coloured, with its lengthening or contracting forms, creating a symphony of visual amusements.

In terms of lighting, it has been present forever first in the form of fire, candles, gas then electric and now solar lamps. It gives the garden a nocturnal magic, transforming it, playing on invented shadows, putting into perspective the vegetation, the movement, the structures. Fairy-like, light can also be a source of messages, of stories told visually, of new customs. It utilizes technology in constant evolution, responding to ecological goals more and more important in terms of energy conservation.

Third accomplice of the garden, the wind has a particular, contradictory place. In response to this invisible and uncontrollable enemy, people have created structures to protect elements from the effects of hydric and physical stress, to limit the way its real force dries the soil and bends the vegetation. But it is easy to forget its function in pollinization and the fact that it facilitates photosynthesis. Wind puts the garden in movement, making it sing, as one can see in the new forms of garden installations that use it in a positive, inventive and poetic way.

呼吸盒子 / breath box - 2014
NAS Architecture :
Hadrien Balalud de Saint Jean,
Guillaume Giraud & Johan Laure (法国)
Festival des Architectures Vives (法国)
Photos ©: Paul Kozlowski / photoarchitecture

这个位于法国拉格兰德莫特的小屋向水平线伸展着，并借助可移动反光模块的作用，与海洋、光和风产生互动。这些模板拼接成一片大型镜面，衍射并转换周围的景观，交错变化的表面吸引着游人的注意，这件海岸上的装置成为不断更新的多感官体验场所。

In La grande-Motte, this pavilion oriented toward the horizon, plays with the sea and its colleagues, the light and the wind, thanks to the involvement of mobile, reflective modules form a large mirror that diffracts and transforms the landscape. Attractive and interactive, this maritime fabrique is the theatre of unceasing multi-sensory experiences.

白色水源
white source
法国 France

2012

在法国的大西洋卢瓦尔省,时间凝固在白色王冠修道院内院中央的一口井中,井水早已被取之殆尽。这件装置作品隐喻着从深处喷涌而出的井水,半径一直延伸到内院周边的老树。它象征性地占据着空间,向世人展示具有代表性的圣像、宗教类的装饰或者有特定用途的物件,比如探测地下水的探测棒。这是一件硕大丰富又极为精确的装置,向这个充满历史意义的精神领地致意。

In the heart of the cloister of the abbey Blanche Couronne, in the department of the Loire-Atlantique, time freezes in the depth of the wells, where water has been drawn thousands and thousands of times. A plastic representation of water springs from the depths and radiates to the periphery of the ancient trees in the cloister. Water symbolically occupies the space and gives birth to icons, religious ornaments and objects with a specific usage, such as a witching stick. A generous installation with a pertinent message, it renders homage to this place, historically laden with spirituality.

Marie-Hélène Richard & Stéphan Bohu
法国
Biennale Estuaire, Communauté de Communes Loire et Sillon
法国

Photos ©: Marie-Hélène Richard

美之园
garden of beauty
法国 France

1998

此花园把光线和植物巧妙地结合在一起，形成一首对水和其无穷形态的赞美诗。一条激流小水渠从黑水池流泻而出，如瀑布般级级下降，而此黑水池则如同镜面一般反射着天空的光彩云影，上面漂浮着一个奇异的立方体。水流沿着立方体的两片洁白墙面悄声无息地垂淌而下，在墙板上勾画出一圈圈的涟漪，呈现出巧妙的透明效果。长条木板在立方体内部铺出一片广场，朝向一排桦树敞开，桦树的另外一边则是种植了茂盛蕨类植物的阶梯式空间。这个由简洁流畅的线条勾画出的避风港为人们提供一个安逸舒适的空间，邀请人们来此放松休息、享受片刻宁静。

Hymn to the world of water and its infinite forms, this garden subtly associates light and vegetation. Water cascades from a succession of metallic pools. It springs from a pool, or rather, from a mirror of black water reflecting the sky. In the middle of the black pool is a large floating cube. On the two pristine sides of this "drift-cube", sparkling and silent little waves flow, playing with their own transparency. A covered deck opens up to a border of silver birches that precedes an amphitheatre of exuberant ferns. Flamboyant touches of plants accentuate the impressionist atmosphere. This soothing harbour with both pure and fluid lines, is an invitation to peace and quiet, to serenity.

Jean-Pierre Delettre, Michèle Elsair, Oxalis
法国&瑞士
L'art du jardin/Clarins, Domaine national de Saint-Cloud
法国

Photos ©: Oxalis

爱丽丝与克拉拉
alice & clara
法国 France

在布厄，夏朗德省的心脏地带，镇上的旧洗衣池带领人们跟随"爱丽丝梦游仙境"进行一次"镜中奇遇"。池边的巨大实心橡木框上装饰着英国栎和山毛榉的叶子，是雕刻家安杰伊·弗罗那和细木工安东尼·克鲁切克的作品，这片庄重的水面邀请人们来此照镜，以开启一段梦境之旅。

In the heart of Bouëx, in Charente, the old lavoir of the village, the place where women went to wash clothes, invites us to glimpse Through the Looking-Glass, the sequel to Alice in Wonderland. Adorned with a giant frame of massive oak, cut from English oak and beech leaves by the sculptor Andrzej Wrona and the cabinet maker Antonni Kruczek, this majestic body of water attracts us and offers a chance to gaze at one's reflection, for a dream-like moment.

Gérard & Marie Denis
法国
Le PAC'Bô, Association ADN
法国

Photos ©: Marie Denis

红灯笼
the red lantern
美国 USA

**Cao|Perrot, Place Studio,
J.P. Paull / Bodega Architecture**
法国&美国
Cornerstone Gardens Festival, Sonoma
美国

Photos ©: Stephen Jerrome

这件景观作品是为了纪念19世纪建造加利福尼亚铁路的中国工人，正是他们开创了今天旧金山的中国城，有史以来亚洲之外最大的华人居住地。作品中的所有元素都让人联想到这些移民的出身：作为象征的灯笼、覆上蛋壳的漆木、水晶吊坠的形状以及颜色。通向灯笼小屋的铁路枕木更为这件作品带来令人惊奇的效果，与位于这个特殊水上公园边缘的巨型筷子相得益彰。

This landscaping installation is a homage to the Chinese people who helped construct the transcontinental railroad from Omaha to California in the 19th century and who founded San Francisco's Chinatown, still harboring the largest Chinese community outside of Asia. Everything here evoques the origins of these immigrants: the choice of the symbol of the lantern, its material and the form of the crystal pampilles, the railroad cross-ties leading to an astonishing fabrications, like giant sticks raised to mark the threshhold of this singular water garden.

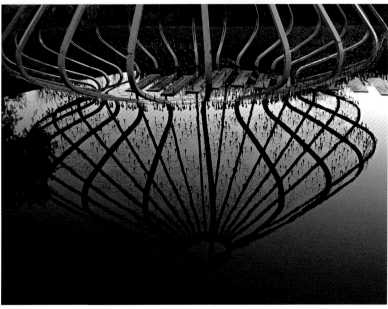

协奏三曲 / 水

漂浮森林
ss ayrfield

澳大利亚 Australia

1972

这艘舰船于1911年建造于英国，曾在第二次世界大战期间于太平洋战场上运送士兵。1972年解除武装并被送到悉尼西部霍姆布什湾的拆除工地。虽然和其他被遗弃的船只一样停泊在海湾里，但只有这条船上才出现令人称奇的茂密植物。一片红树林——典型的海洋湿地森林——在此自发地生长，使它变成一座"漂浮森林"，并以此别名广为人知。

Constructed in England in 1911, this ship for transporting troops served during the Second World War in the Pacific. It was disarmed in 1972 and sent to the demolition shipyard of Homebush Bay, west of Sydney. Like certain other boats, it lies abandoned but it is the only one where nature has astonishing reclaimed her rights. A mangrove, a forest typical of maritime marshes, grew spontaneously, transforming it into a "floating island", the nickname by which it is known today.

Photos ©: Andy Brill p.88
& Steve Dorman p.89

维纳斯与偶然
the game of venus and chance
法国 France

2006

在一个传统的水上花园中，在浮桥、水池以及茂盛植物的衬托下，波提切利的"维纳斯的诞生"获得了崭新的演绎。爱与美的女神所站立的巨大贝壳犹如一个河岸岬角，由此可看到一个巨型拼图的碎片在水面上慢慢漂流。

这里是胸部、嘴唇，那里是眼睛、手掌，这些片段以记忆游戏的方式让人们重新认识维纳斯，还有准备给维纳斯披上斗篷的仙女，以及吹着西风把维纳斯吹向岸边的风神奇菲尔。

Inside a classic aquatic garden with pontoon, pond and lush vegetation, Botticelli's Birth of Venus is revisited. The ample shell from which rises the goddess of Love and Beauty forms a promontory where one can observe the pieces of a giant jigsaw puzzle floating slowly to the surface of the water.

Here, a breast, a mouth, there the eyes, a hand, allow us by the game of memory to recognize Venus, as well as the nymph ready to envelop her in a protective veil, but also Zephyr, the wind of the west who, with his breath, drives her to the shore.

Maro Avrabou & Dimitri Xenakis
希腊&法国
Festival International des Jardins de Chaumont-sur-Loire
法国

Photos ©: Dimitri Xenakis

大量的水生植物和沼泽里的小居民（青蛙以及北螈）诗意般的存在，使这件作品生动起来，水波不停地试图将画面的碎片组合起来，描画出梦想中大自然的情爱故事。

A work that lives through many water plants as well as through the poetic presence of the little inhabitants of the marsh (frogs, newts...), the waves play endlessly at assembling the fragments of the scene, sketching the erotic geography of an ideal nature.

漂浮岛
floating island
法国 France

2001

巴斯克地区的贝尔古埃-维耶勒纳夫村庄在比杜兹河上举办了一场有趣的晚宴。菜单中包括了"塘汤"、"酿鸭"、"蛙腿"等菜色,当然还有与项目同名的甜点,所有菜肴都在邻近的罗马石桥的塔楼上烹饪而成。晚餐结束后,半沉浸在水中的宾客可以继续围坐在木筏桌旁,随波逐流,享受夜晚时光。

A strange feast took place one night in the water of the Bidouze, a tributary of the Adour River in the tiny Basque village of Bergouey-Viellenave. On the menu, "Pond Soup", "Stuffed Duck", "Frog Legs" and of course the eponymous dessert ("Ile flottante", a French standard consisting of meringue floating in vanilla custard). The entire meal was cooked on the tower of the nearby Roman bridge. At the end of dinner, the semi-submerged invitees continued the evening by diving and playing with the table-rafts in the current.

Doug Fitch & Mimi Oka
美国

Photos ©: Doug Fitch & Mimi Oka

这些漂流在水面上的圆桌为"生存艺术家"道格·菲奇和米米·奥卡的设计作品,是他们所组织的"俄尔浦斯节庆"的一部分。他们探索食物的世界,以餐宴的方式为人们带来超越时空的多重感官体验。

This tableau-vivant, orchestrated by Doug Fitch and Mimi Oka, who are both "subsistence artists", is part of the "Orphic feasts" that they produce, exploring the world of the comestible through multi-sensorial banquets from another time.

自由荷叶
lotus in motion
加拿大&法国 Canada & France

2011

无论在现代的水池中还是在历史性花园中——如温哥华的凡杜森植物园，这些巨大的荷花叶片都打破了场所的常态而带来新奇效果。

这些抽象画面从克罗德·莫奈所绘的睡莲获得灵感，也受到"跑冰排"时漂流在加拿大河川上的浮冰的启发。

Whether they take place in a contemporary pond or in a historic garden, like the Van Dusen Botanical Garden in Vancouver, these immense lotus leaves disturb the ordering of spaces.

These abstract paintings were inspired as much by the nymphs painted by Claude Monet, as by the blocks of ice racing along Canadian rivers during an ice flow.

Gordon Halloran
加拿大
Jardin botanique Van Dusen et le jardin chinois classique Sun Yat Sen de Vancouver
加拿大
Jardin des Tuileries de Paris
法国

Photos ©: Gordon Halloran

所有荷花叶片各个相异,它们在水面上滑动、原地转圈,以不同的颜色与透明度相互重叠,在阳光照射下,与水底的斑烂光影相映成趣。

All different, they slide, turning on themselves and playing in the water, with their superimpositions of colours and their transparency, all interactive.

浮游植物
phytoplankton
法国 France

2011

在艺术作品集《濒临危险的空气》中，平川滋子向各种自然元素表达敬意，它们维持着这个星球逐趋脆弱的平衡关系。在莫尔比昂省的一个布列塔尼湖边，她以造型艺术的表达方式重新制作出蓝藻，一种最古老的生命形式。平川滋子希望借此提醒人们关注数以万计的藻类植物和浮游生物，它们通过光合作用产生赖以生存的能源，同时在海洋、湖泊以及河流中释放出地球上65%的氧气。

In one of her artistic installations, "Air in Peril", Shigeko Hirakawa renders homage to the different natural elements that contribute to the balance of our planet, now fragile. On a Breton lake of Morbihan, she presents sculpted versions of cyanobacteria, one of the oldest forms of life. Shigeko Hirakawa attracts our attention to the tens of thousands of algae and plankton that use photosynthesis to create the energy they need. And in seas, lakes and rivers, they produce 65% of the oxygen of our planet!

Shigeko Hirakawa
日本&法国

Photos ©: Shigeko Hirakawa

潮汐花
tide flowers
美国 USA

Flower opened up during high tide

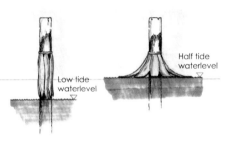

Low tide waterlevel　　Half tide waterlevel

Stacy Levy
美国
Hudson River Park Trust
美国
Photos ©: Stacy Levy

大自然拥有一定的节律，这一点总是被人们遗忘，甚至视而不见。而这却正是纽约哈德逊河的情况，河面的高度每天都随着潮汐时间的不同而起伏。一些巨型的玫瑰色和橘红色花朵被固定在用来减弱水流强度的木桩上，它们在涨潮时绽放，落潮时闭合，见证着大自然的变化。它们也引起相邻公园内游人的注意，借此意识到太阳与月球的万有引力共同作用下所产生的效应。

Often forgotten by men, nature does not change its rhythm even if we no longer perceive it. That's the case for the Hudson River, in New York, its level fluctuating each day according to the tides. Highlighting this flux and reflux, big red and orange flowers are attached to the piers meant to break the force of the movement of water. They mark the river's variations by blossoming at high tide and closing up at low tide, recalling to the people who frequent the nearby park the combined effect of the force of gravity of the moon and the sun...

水上迷宫
labyrinth on water
法国 France

2009

造型艺术家弗郎斯·德·郎尚很早就为自己的艺术创作选择了一个特殊的主题：迷宫。1995年，她甚至发明了"迷宫师"这个词来称呼自己。

无论她赋予迷宫什么形式，绘画或者是装置作品，弗郎斯·德·郎尚在创作时总是认为：公众在为了解决迷题而流连往返的过程中，几乎不知不觉地便吸收了一部分最初设计中的艺术含量。

在昂古莱姆，她把一些小径和死胡同铺展在夏朗德河的一条可通航的支流上，这个水上迷宫由覆盖着荧光玫瑰红色绒毯的木托盘拼组而成，构成一朵巨大的几何形睡莲，很快就吸引了黑水鸡来此搭窝。

A mixed-media artist, France de Ranchin chose very early a particular theme for the ensemble of her artistic work, the labyrinth. In 1995, she invented the term "labyrinthist" to describe herself.

Whatever form she gives them, paintings or installations, France de Ranchin conceives them by thinking that the public, in lingering to resolve the puzzle, absorbs, without realising, a little of the artistic inspiration of the initial design.

In Angoulême, on a navigable branch of the Charente, de Ranchin unfurls her paths, dead-ends and cul-de-sacs, made of wood pallets covered in flourescent pink carpet that form a giant, geometric water lily, where a little pool of water does not take long to form.

France de Ranchin
法国
ACAPA & École d'Art du Grand Angoulême
法国

Photos ©: France de Ranchin

偏离常道
off the beaten path
法国 France

2012

奥雷莉·巴尔贝和劳拉·胡戈鲁以独特的方式描绘了里昂金头公园的城市环境，创造出一种看似荒谬却又饶富诗意的形式，一段看起来极不可信的行人穿越道。这些由柳条编织并缀上了树叶的地面标识显得相当脆弱，它们铺陈于水面之上，继而飞入树林之中。这条无法通过的行人穿越道邀请游客进行一次精神之旅，他们可以停留在水边冥思，遐想自己在水面上行走或者沿着一条只有自己才知道终点的小路前行。

Interpreting in their own way the urban environment of the Park of the Tête d'or, in Lyon, Aurélie Barbey and Laura Ruccolo invent an absurd and poetic form of improbable pedestrian paths. Woven of willows and embroidered with leaves, the path is at first fragile marks on the soil, then it rests on water and continues by floating off into the trees, impassable. The visitors are invited to a mental experiment. And if they have ever dreamt of walking on water or of a path leading away, known only to a few, they stay, contemplative, on the bank.

Aurélie Barbey & Laura Ruccolo
法国
**Fêtes des Feuilles,
Parc de la Tête d'or**
法国

**Photos ©: Aurélie Barbey
& Laura Ruccolo**

协奏三曲 / 水 103

水上印记 1
water footprint 1
法国 France

2014

一座安置在泰瓦黑庄园中的过桥被奇怪地分成两半，中间支撑它们的是一个巨大的水袋，柔软且会晃动，人们必须小心谨慎地光脚走过，并在它的表面留下暂时的印记。

这件互动式装置作品同时也是一个具有改革理念的方案，水袋的容积有55立方米，正是每个法国人的年耗水量，这个数值不包括重要的消耗产品中所包含的或者需要的水量。这样一种身体的和感官上的穿行，把这个法国庄园中的一部分个体"生态印记"呈现出来，并且让人们意识到这个国家在用水提供上并不能自给自足，然而他们的需求却超出了世界平均的30%！

A bridge, installed in the Domaine de Trévarez, is astonishingly cut in two. In the middle of its path is an enormous cistern, a supple and shifting sack of water which one must cross barefoot, carefully, marking its surface with fleeting footprints.

This interactive installation is also an activist project, the reservoir holding 55 m³, representing the volume of water consumed by each French person yearly, not counting products imported and consumed that need or contain water. This physical and sensory crossing permits us to visualise a part of the individual "ecological footprint" and to learn that France is not autonomous in this area, as its needs are 30% higher than the world average!

Shigeko Hirakawa
日本&法国
Chemins du patrimoine en Finistère, Domaine de Trévarez
法国

Photos ©: Shigeko Hirakawa

某生存者心中的生活绝境
the impossibility of life in the mind of someone living

德国 Germany

在魏玛市中心，一件奇异的物体占据着阿斯巴赫桥下的位置：一个在夜幕降临的时候会被灯光照亮的大水族箱，五条金鱼畅游其中。这件吸引人的装置作品并非来自偶然的灵感，它是这个前东德城市街区的居民和工人的一种生活映射，人们每日在危机中生存却没有太多的选择余地。

A strange object has taken up space under the Asbach bridge, in the centre of Weimar: a big aquarium, lit at nightfall, where five goldfish swim tranquilly. But this seductive installation is not innocent, it is a metaphor of the life of the inhabitants and workers of this urban quarter of former East Germany where, without many alternatives, the crisis is lived daily.

Holger Beisitzer
德国
Ville de Weimar
德国

Photos ©: Holger Beisitzer

南特<>圣纳泽尔河口
nantes <> saint-nazaire estuary

法国 France

Jean-Luc Courcoult
法国
Erwin Wurm
奥地利
Huang Yong Ping
中国
Le Voyage à Nantes
法国

Photos ©: Bernard Renoux / LVAN

这是一项从2007年开始展开的河流和土地尺度的非凡艺术探索之旅，到如今，已经有28件作品被分散设置在卢瓦尔河畔，位于南特和圣纳泽尔之间60千米长的河口地带。这是一条真正的阿里亚娜之线（源自古希腊神话，用来比喻解决复杂问题的线索），这条点缀着艺术作品的路线连接了河岸边的不同城镇，全年开放免费参观，让人们得以发现一些不同寻常的出色场所，它们见证了河岸工业景观和自然景观的持续变化。沿着河水行进，人们会在库埃龙发现让－吕克·库尔库勒的"卢瓦尔河中的房子"，它的基础被淤泥所固定并稍稍沉陷，犹如一个建筑残骸。

In 2007, an uncommon artistic adventure began, on the scale of a river and a territory. Today, a collection of 28 permanent works, produced in situ, dots the banks of the Loire, along the 60 kilometres of its estuary between Nantes and Saint-Nazaire. Like the spider's thread that led Theseus out of the labyrinth, this itinerary, accessible all year long, allows one to discover atypical, remarkable places, in the different riverside villages, testifying to the incessant change of its industrial or natural landscapes. Following the current, one encounters in particular, near the town of Couëron "The House in the Loire" of Jean-Luc Courcoult, its foundations stuck in silt and sludge, leaning slightly like a shipwreck.

在勒佩勒兰，马帝尼耶运河边，雕塑家欧文·武尔姆安置了作品"不可思议"，一条倾斜且弯曲的帆船，好像正在被一股不可抗拒的力量拉向水中，把这个平凡的场所瞬间转变成一个看似荒诞的世界。

在圣布雷万莱潘、位于河流与海洋之间的河口地带，一个巨型海洋怪物的骨架出现在沙滩上，仿佛活生生的动物。

At Le Pellerin, at the edge of the Martinière Canal, the sculptor Erwin Wurm installed "Misconceivable Mnéol". 'Mécocevable'N", a sailboat that leans and folds, as if irresistibly drawn by the water, making this ordinary place tip into an absurd universe.

And at Saint-Brévin-les-Pins, located on the estuary between the river and the sea, the skeleton of an immense sea monster rises up on the beach.

艺术家黄永砅的这件作品起名为"海蛇",在水中时隐时现的脊柱让人们联想起远处圣纳泽尔桥的曲线,而其脊椎骨的形状也令人想起一种当地典型的建筑结构——那些点缀在大西洋海岸、架空在水面上的钓鱼木屋。

Entitled "Sea Serpent" by the artist Huang Yong Ping, it seems alive, its vertebral column playing with the weather and evoking the curve of the bridge of Saint-Nazaire that we perceive from afar. The form of the vertebrae also recalls an architectural element characteristic of the region, the fishing huts on stilts that punctuate the Atlantic Coast.

薄雾雕塑
standing cloud
法国 France

2013

瓜卢公园的桦树林中存在着一件非同寻常却又富有诗意的雕塑作品，改换了景观的面貌。每隔一段时间一片云团就会慢慢形成并开始把游客包围起来，轻盈的薄雾贴在皮肤上则变化成飘动的细微水滴。大自然导演出这片梦境般的薄雾景色，空气、光与影都参与其中，随着时间的推移将云雾塑成不同形态。

In the Park of Goualoup, a grove of birches host a strange and poetic sculpture, transforming the landscape. A cloud Intermittently forms and envelopes the visitors with a light mist, leaving on their skin fine, fleeting drops. This dreamlike fog, created by nature, plays with air as well as with shadow and light, and is sculpted by them as the hours pass.

Fujiko Nakaya
法国
Domaine de Chaumont-sur-Loire
法国

Photos ©: Éric Dufour
p.110, p.111 左下/bottom left
Sophie Barbaux
p.111 左上/top left & 右/right

白云
bai yun - white cloud
美国 USA

2011

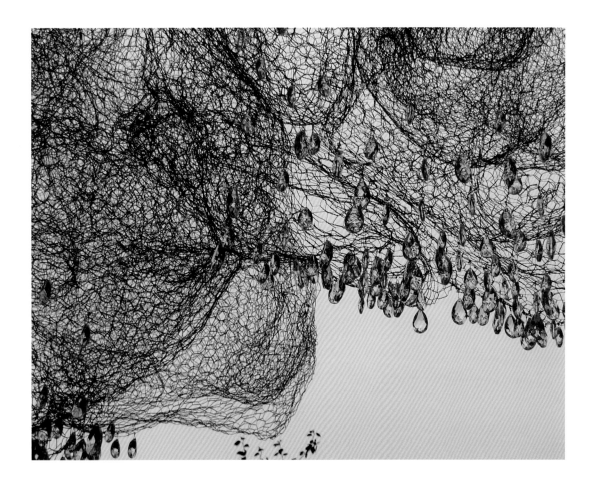

这个花园的名字"白云",意指"白色的云朵"。用清晰准确的线条勾画轮廓是14世纪日本浮世绘的特征,花园的设计受此启发,以细密的网格层层堆叠出积云的形状,上面挂满五千条水晶流苏。云朵之下庇护着由细沙、碎牡蛎壳和回收玻璃堆积成的小丘。从黎明到月光初上,这个微型景观与周围的植物、风和光纤不断地产生互动。

Entitled Bai Yun, "white cloud" in Vietnamese, this garden finds inspiration in the Japanese prints of the 19th century, with its lines precisely cut. Cumulus clouds sculpted in multiple layers of fine mesh and adorned with five thousand cut crystal beads, overhang sand dunes as if sheltering these mounds of crushed oyster shells and recycled glass. This micro-landscape interacts as much with the different plant environments surrounding it as with the blowing wind and the light, as with the sunrise and moonlight.

Cao|Perrot,
美国&法国
Cornestone Gardens Festival, Sonoma
美国

Photos ©: Stephen Jerrome

回到源头
back to the source
法国 France

2010

这件景观作品呈现出亚瑟王传说中仙女薇薇安与魔法师梅林在一个奇怪的喷泉边第一次相遇的情景。奥雷莉·巴尔贝和劳拉·胡戈鲁在这个位于布劳赛良德森林边缘的伊坊迪克公共洗衣房前重新诠释这个情境，以轻盈通透的巨型气泡象征性地呈现了此传说故事神奇与美妙的特质，也使此地的往昔生活仿佛苏醒了过来。

The first encounter of the fairy Viviane and Merlin the Magician, on the edge of a strange fountain, is revisited by Aurélie Barbey and Laura Ruccolo in front of the lavoir, the traditional washing place of the town of Iffendic, bordering the forest of Brocéliande. The magic and marvelous character of the legend is symbolized by giant bubbles, with a light and vaporous presence, bubbles that also recall the everyday use of this site in the past.

Aurélie Barbey & Laura Ruccolo
法国
Festival Étangs d'Art de Bretagne
法国

Photos ©:
Aurélie Barbey & Laura Ruccolo
p.114 右/right & p.115
Boigerninie
p.114 左/left

交响田园
harmonic field
法国 France

2010

这条位于法国马赛的交响乐之路把500件乐器请上大自然舞台，它们以令人称奇的方式栖息在高处，构成一件件雕塑般优美的声学装置。设计师的灵感来自巴厘岛农民的风玲稻草人——其特点在于随着风的节奏而发出不断变化的自然之声，此作品同时也效仿码头船只上那些绳索或者挂钩有节奏地摇摆汇合而成的断断续续的旋律。

彩色梯皮、震动的鼓、竹哨以及螺旋桨鸣音器在与空间对话的同时，也自然产生共鸣，形成美妙和声。钟琴、声弓以及旋转的音乐盒捕捉流动的空气，既是音乐家又是作曲家。

Landscaped, this philharmonic path plays host to 500 instruments in Marseille, all astonishingly perched, forming an acoustic and sculptural installation. Its sources of inspiration are the wind-blown scarecrows of Balinese peasants, having the particularity of emitting a natural song created by the cadence of the wind. The installation is also inspired by the broken melodies and swaying rhythms of the riggings and halyards of boats in port.

Dialoguing with space, resonance and temporality, such strange objects as chromatic tepees, vibrating drums, but also bamboo pipes and propeller-sirens offer their spontaneous harmonies. Windmil-powered glockenspiels, sonorous bows and spinning music boxes capture the air in movement, musician and composer at once. Confronted with these musical

Pierre Sauvageot
法国
Lieux Publics
法国

Photos ©: Vincent Lucas

面对这些由谐波大提琴、五音阶圣杯、长笛树和响尾蛇钓鱼杆组成的音乐森林，表演艺术家们引领参观者来到一曲精心策划的最终乐章，邀请人们闭眼聆听或者仰天沉思，联想生命中的影象，也捕捉记忆中的故事……

随着缓缓前进的漫步，观众们在风的符号中徜徉旅行。这个交响乐装置提供了人们一场独特的经验，借此向作曲的概念以及人们作曲时日益依赖的科技方法提出质疑。

forests of harmonic cellos, of pentatonic treasures, of trees of flutes and side-winding fishing poles, performance artists guide visitors to an orchestrated finale. They invite one to listen, either with eyes closed, or in the contemplation of the sky, evoking images, capturing tales…

Throughout the course of their walk spectators travel in the alphabet of the winds. The experience offered by this symphonic apparatus questions the notion of composition and investigates, with relevance, the technological methods that are more and more associated with it.

花朵 2.0
flowers 2.0
法国 France

2014

格勒诺布尔市植物园内的两个大草坪在几天之内就变成了一片布满鲜花的草甸。成千个塑料瓶做成的上百个五颜六色的花朵完全改变了这些草地的面貌。

人们在其中闲逛，好像在私人迷宫中漫步，随心所欲地创造新的路线。这个充满诗意的装置作品邀请人们在草地上的彩色阴影下停留休息。当人们躺在其中，这个与众不同的空中花园中布满了透明的花朵，其中一些看起来像是巨大的蜻蜓，纤细而优美。花朵在微风的轻拂下，微微摆动，栩栩如生，而花丛中则流淌出一曲曲无名的旋律。

A joyous flowered meadow grew for a few days on two grassy areas of the Jardin des Plantes in Grenoble. Hundreds of multi-colored flowers, made from thousands of plastic bottles, transformed the space.

People stroll there as in a personal labyrinth where they might create paths at will. This poetic installation is an open invitation to take a pause, under the colored shadows written on the grass. And, when one lies down, an unusual garden of sky appears, full of transparent flowers, some of which seemed transformed into slender, giant dragonflies. When the breeze rises up, it passes through the flowers that move lively in response, letting escape an unknown melody. The music is doubly interactive.

Pierre Estève
法国
**Festival les Détours de Babel /
La Métro à Grenoble**
法国
Shooting Star
法国

Photos ©: Pierre Estève

这些风之歌具有与游客互动的特质,茎秆一被摇动,便可以改变韵律。而艺术家皮埃尔·埃斯塔夫的一首音乐作品又对此装置进行了补充,每当有游客穿越其间就会奏响这首旋律。

这些花朵装置由大人和小孩共同组成的"花匠学徒"团队所制作。这件具有参与互动特性的作品创造出一种混合的自然,一种崭新的生物多样性,是消费社会的成果。这个当代性十足的生态系统作品已经开始出现在不同的场地中,自然的、景观的以及城市的,它也可以改变尺度,甚至变成一片可持续发展的或者充满光亮的有声森林,意喻着这个世界的变化。

By playing with the stems of the flowers, people can modify this wind song. Meanwhile, the motion of passerbys starts the diffusion of a musical composition by Pierre Estève.

Produced by different groups of "apprentice florists", young and old, this participatory installation creates a hybrid sort of nature, a new diversity, fruit of the consumer society. The project of contemporary ecosystems was started in different spaces, natural, urban and landscaped. It can change scale, become a forest that is sustainable and sonorous, maybe even luminous, a metaphor of the mutations of our world.

河岸
riverine
日本 Japan

2009

在阿贺野河边的一片泛洪平原上，种植了600棵高度超过5米的竹子。它们的顶端是聚苯乙烯材质的彩色球，让人们联想起巨型芦苇的花序。这些庞大草本植物的柔软茎秆创造出独特的舞步，随着光和风的变化而摆动。

六个月之后，同样柔软的真实柳树取代了这些竹子，以相同的布局方式种植，并且创造出一片崭新的演进式动态花园。

On the banks of the Agano river, 600 bamboo pool more than 5 metres high were planted in a flood plain. At the top, coloured styrofoam balls evoke the inflorescence of flowers of giant reeds. Like immense grasses, these stalks of supple bearing create their own choreography through the variations of light and wind.

At the end of six months, living willows, just as flexible, were installed in their place, taking up the same geography of plantation and creating a new garden in movement.

Stacy Levy
美国
Niigata Water & Land Art Festival
日本

Photos ©: Stacy Levy

万有引力2号
gravity #2
法国 France

2008

在第戎市中心达西广场的纪尧姆拱门下，成百上千个白点在一片铺着草皮的园圃上形成了一个从远处看来令人称奇的图案。这片白色花海是一件临时的装置作品，人们走近它之后才发现原来是固定在纤细枝干上的3600个乒乓球，在微风的轻拂下摇摆。黄昏时分，这些洁白无瑕、花形极为规律的花朵借助一个"黑色灯光"的设置完全变换了面貌，不仅使空间产生变化，也强化出城市的透视轴线。

In Dijon's Place Darcy, under the Porte Guillaume, a multitude of white dots form a surprising graphic network in the distance. As one approaches the large lawn, planted with flowers for the installation, one starts to distinguish the 3 600 ping-pong balls stuck to thin stems that wave in the breeze. At nightfall, these virginal blossoms, with their astonishingly regular bearing, metamorphose by means of a "black light" device, transforming the space, accentuating the created perspective.

Dimitri Xenakis
法国
Jardins éphémères
法国

Photos ©: Dimitri Xenakis

空间阴影
spatial shadows
法国 France

Thomas Klug
法国&巴西
Désert de Retz
法国

Photos ©: Thomas Klug

由德蒙维尔先生所建立的"雷茨荒漠"诞生于法国大革命前夕，它也被称作"光明花园"，这个英中式的园林包含了17个标志性的建筑物，其中最主要的一个叫做"被摧毁的柱子"，整个庄园以它为中心而组织。为了照亮这栋独特的建筑物，一个发光的球体在十几米直径的圆形轨迹上围绕建筑作圆周运动，为建筑立面和它周围的环境中营造出时而阴影时而逆光的微妙效果。

Le Désert de Retz was created on the eve of the French Revolution by Monsieur de Monville. Also called "The Garden of Lights" it includes 17 fabriques (small buildings or ruins) emblematic of Anglo-Chinese gardens. The property is organized around a fabrique of about fifteen meters in diameter, the "ruined column". To light this singular structure, a luminous globe in circular movement organizes the spatial relations of the facade and its surroundings by a subtle game of shadows and backlighting.

欲望和威胁
desire and threat
瑞士 *Switzerland*

2012年冬天，两只巨大的鸟栖息于日内瓦喷泉大街边的树枝上。轻盈的金属网纹理使它们形体透明，即便有4米长，在白天也并不引人注意，甚至几乎与树枝融为一体。夜晚降临时，它们发出光亮，明显地展现出它们怪异的存在，仿佛两个静止不动的监视者观察着行人和周围的建筑。紧张、捕食、诗意与神秘同时附着在这件巧妙而又脆弱的装置作品中，它把人类重新放置到物种的尺度上……

During the winter of 2012, two gigantic birds perched for a moment on trees overhanging the rue de La Fontaine, in Geneva. Discreet in the daytime, their light wire mesh bodies, four metres long, almost camouflaged them in the branches which played with their transparency. At nightfall the birds lit up, brightly announcing their strange presence. Immobile guardians observed the passers-by and the surrounding buidlings. Tension and predation, but also poetry and mystery inhabited this subtle installation that returned man on a level with other species, equally vulnerable.

Cédric Le Borgne
法国
Festival Arbres en Lumières
瑞士

Photos ©: Cédric Le Borgne

镜屋
mirror house
丹麦 Denmark

2011

在哥本哈根的中央公园中，一个有趣的小房子取代了遍布涂鸦的旧房子，为孩子们带来新的游戏空间。老房子的墙面和屋顶都以环保的方式重新用高热干燥木材修饰一新。山墙和百叶窗则别出心裁地使用光面不锈钢作为饰面材料，整个墙体变成了一面镜子，根据不同曲面而呈现出变形的效果。这片巨大的反射表面同时吸引着大人与小孩，他们与镜子里自己的影像互动，背景中的公园与树木亦被投影其中，换化为另一种景观。

A strange pavilion has replaced an old, grafittied building, and has become a place for children's activities in Copenhagen's Central Park. It was rehabilitated in an ecological way, using thermo-heated wood for the facades and the roof. For the gables and shutters, an original material was chosen: polished stainless steel, a "fun mirror" that distorts as it curves. Immense, these reflecting and interacting surfaces attract young and old alike, allowing them to play with their image, with the trees of the park as a backdrop, for they, too, have been transformed by the mirrors.

MLRP
美国&丹麦
Copenhague Municipality, CAU & Mærsk McKinney Møller Fond & wife Chastine McKinney Møller
丹麦

Photos ©: Stamers Kontor

眩晕
vertigo
法国 France

2012

以萨尔特省凡尔耐伊-勒-塞提夫的芒日城堡为蓝本，"爱丽丝"项目中的"眩晕"被选取出来，呼应着这个地区铭刻在历史上的神秘事件以及不协调而又奇怪的建筑加建。此装置作品出现在临近城堡的贝尔赛森林中，镜面圆盘使得的树木显得更为高耸，也将人们的视线延伸到地球中心。景观的扭曲让人失去参考依据，转而徘徊于梦境与现实之间、在凡尔耐伊与种种奇观之间……

Just like its setting, the château de Mangé at Verneil-le-Chétif in the French department of La Sarthe, the project "Alice" of which "Vertigo" was a part, echoed the mysteries that punctuated both the history of this place and the architecture, with its sundry strange additions. In Vertigo, mirror disks extended the verticality of the trees of the nearby state forest of Bercé, drawing the eye to the heart of the earth. This distortion of landscape led to a kind of disorientation between dream and reality, between land and Wonderland…

Gaëlle Villedary
法国
Pays Vallée du Loir & Saxifrage
法国

Photos ©: Gaëlle Villedary

明晰
clear cut
瑞典 *Sweden*

Joakim Kaminsky & Maria Poll
瑞典
Forêt de Medelpad
瑞典
Photos ©:
Kjellgren Kaminsky Architecture

2011年夏天,约阿基姆·卡明斯基和玛丽亚·波尔深入到梅代尔帕德森林中,如同瑞典其他大量林地资源一样,这也是一片可以为私人业主带来收益的松林。他们两人把15米长的布条与镜面材料粘合在一起,为树干缠上一条条窄小的反光带,让人联想起成年的树木在被砍伐前的标识记号。他们在短暂的时间内将某些树干神圣化,借此向这块基地的祖传精神以及不断持续生长的一代代松树表达敬意。这项装置创作刻意只保留一天一夜,是生命周期的写照,永远都在重新开始中。

During the summer of 2011, Joakim Kaminsky and Maria Poll went into the depths of the Medelpad forests, an area commercially planted and harvested, like many forests in Sweden. With fifteen metres of mirror-coated fabrics, they wrapped a reflecting band around many trees, thus recalling the marking that precedes the felling of full-grown specimens. By sanctifying some trunks in an ephemeral way, the artists render homage to the ancestral temporality of the site and the successive generations of pines. An installation that purposely lasts one day and a night, imitating the cycle of life, in perpetual rebirth.

普赛克之镜
psyche 法国 France

在威尼斯圣洛克学院里，参观者借助一些便携镜子的辅助得以更好地欣赏丁托列托于16世纪画的天顶画。玛丽·德尼受此启发，为戈纳斯双年展设计了一面可动式大型圆镜，照映出附近戴高乐机场起降飞机所形成的空中芭蕾。

这面镜子在成为沙马朗德省级公园的收藏品之后，折射出另外一种景观：当然有天空，还有茂密的叶簇和建筑物，以及著名的"飞瀑"。随着参观者的欲望，此镜面把周围的环境转换成无穷无尽的活动画面。

At the Scuola Grande di San Rocco in Venice, hand mirrors permit one to better view the ceilings painted by Tintoretto in the 16th century. Drawing inspiration from this practice, Marie Denis conceived the immense circular cheval mirror, mobile as it should be, for the Biennale of Gonesse. It reflected the aerial ballet of the planes from the neighboring airport, Roissy-Charles de Gaulle.

Now become part of the permenant the collection of the Domaine Départemental de Chamarande, the Psyche reflects another landscape: the sky, clearly, but also the foliage and the architecture, like the famous "Buffet d'eau" or water fountains. It transforms, always at will, its environment into an inexhaustible animated painting.

Marie Denis
法国
Collection du Domaine départemental de Chamarande
法国

Photos ©: Marie Denis

像素 pixel
西班牙 Spain

Travesias de Luz
西班牙
La Noche de Luna Llena, Segovia Cultura Habitada
西班牙

Photos ©: Tania Vegazo / Travesias de Luz

2011年圣诞节，光之旅协会在马德里的大街上摆满了特殊的蜡烛，透明的一次性塑料杯里装上水就变成一个个独立的小光源。2012年，协会成员重新利用这种光源，并把它们集中在塞戈维亚城市中的一处开敞空间中。每个杯子被当作一个像素点，当足够数量的点聚集在一起，便能够把一张小幅的设计图转化为一件大尺度的亮明作品。此创作的目的在于向人们展示制作过程，也邀请他们共享夜晚带来的灵感。

On Christmas 2011, the collective Travesias de Luz sows in the streets of Madrid unusual votives, simple clear plastic cups, full of water and furnished with little autonomous indicator lights. With Pixel, in Segovia in 2012, the members of the collective take up the same principle of lighting and concentrate their project on the only free space in the city. Starting from a design of modest size, they use glass as pixel points, or modules that permit, when multiplied, the realization of a large-scale luminous installation. Their proposal aims both to shine light on the production process and to invite people to the illumination of one night.

光陷阱 trap light
荷兰 Netherlands
2011-2012

这些灯笼状的瓶子采用了一种彻底革新的照明技术，它们变换着颜色，从透明转为一系列深浅不同的蓝色调，从黑暗过渡到光明然后又暗淡下来。威尼斯潟湖上的穆拉诺岛以玻璃制品闻名，这些玻璃瓶就是采用传统玻璃制造技艺而制成，并且将光致发光材料融入玻璃之中。它们能够捕获其他光源在空气中释放出的能量，并将其转换成亮光，此光线又再度释放出能量，不断地重新充满玻璃瓶。这种创新方法为花园的环保照明开辟了广阔的前景。

In a design conceived with a radically innovative approach to lighting, strange bottles in the shape of lanterns change colours, going from transparency to a plaid of blues, from shadow to light, and back to shadow again. By utilising the ancestral knowledge of Murano glass-blowers, the artists have added photo-luminescent pigments into the glass itself. These pigments capture the energy diffused in the air by other light sources, convert it into light, which in turn gives back energy capable of being charged again. This process opens many ecological possibilities for garden lighting.

Mike Thompson
英国&荷兰
Gionatta Gatto
意大利&荷兰
Transnatural Art & Design Label
荷兰

Photos ©:
Transnatural Art &Design Label

水母
jellyfish
瑞士 Switzerland

Paradedesign - Ghislaine Coudert
法国
Festival Lausanne Lumières
瑞士

Photos ©: Paradedesign

在卢塞恩的市中心，一些奇特的动物悬挂在欧洲广场的巨大拱门下。这些巨大水母在白天显得洁白无瑕，在夜幕降临时则开始换上不同颜色的盛装，直到天明。它们在空中组成了一个动态的动物寓言集，随着一个隐藏式通风机制的韵律而生动地摇摆起来。这些披着光辉的悬挂装置吸引着行人，驻足欣赏它们自由而又有趣味的曼妙舞蹈。

In the heart of Lausanne, odd animals curled up under the monumental arches of the Place de l'Europe. Pure white during the day, giant jellyfish adorn themselves with changing colours from nightfall until sunrise. Forming a mobile and aerial bestiary, in rhythm with the invisible ventilation that gives them life, these suspensions in garments of light offers to passers-by their free, playful and fairy-like choreography.

禁区 zone
德国 Germany

2011

在一片人迹罕至的丛林中，耸立着一些半埋在土中的窗户，挂着旧窗帘，还能看到一些生活用品，让人联想起一个被废弃的村庄或者一块墓地。这件装置作品是为了一场在德国加尔托举行、纪念切尔诺贝利核泄露事故25年周年的展览而设计的，提醒人们不要忘记1986年4月26日的那场悲惨灾难。夜幕降临后，这个"禁区"窗户内的灯光会点亮，影射着那些被严重污染的土地，或多或少地禁止靠近，然而，尽管暴露于此处的风险已经可以明确地计量出来，某些土地却仍旧有人居住！

In an inaccessible thicket, a constellation of half-buried windows rises, with their old curtains and a few everyday objects appearing through them. Recalling an abandoned village or old cemetery, this installation was conceived for an exhibition organised in Germany, in Gartow, to keep people from forgetting, 25 years later, the nuclear catastrophe at Chernobyl, which took place April 26, 1986. With lights shining thorugh the windows at nightfall, Zone evokes heavily contaminated areas, more or less forbidden, where despite the quantifiable risks, the lands are certainly still inhabited.

Cornelia Konrads
德国
Westwendischer Kunstverein
德国

Photos ©: Cornelia Konrads

温室效应
greenhouse effect

2005-2014

法国, 斯洛文尼亚, 葡萄牙, 英国, 新加坡 France, Slovenia, Portugal, United Kingdom, Singapore

这个从汽车到温室的转型使此作品具有幽默盗用的特质，这些"媚俗"的温室中塞满了被彩色日光灯管照亮的人造植物。但创作者正借助了这份由于无功能而产生的空虚，来向汽车在日常生活中不可或缺的存在提出质疑。汽车对城市和乡村的入侵，以及作为地球上最确定的污染源的角色已经得到证实，但目前我们只能认识到这些问题，却没有办法完全脱离汽车！

Humour took the day in this installation that transformed automobiles into greenhouses, kitschy greenhouses saturated with artificial plants and lit by coloured fluorescent tubes. The vacuity born of the non-functionality of these structures shone light upon the inescapable presence of cars in our daily lives. Their invasion of cities and countrysides, as well as their much too efficient pollution of the planet have been proven. One cannot help but notice, however, that we seem unable, for the present, to do without them!

Maro Avrabou & Dimitri Xenakis
希腊&法国

Superflux / Roger Tator
法国, 2005

Nuit Blanche de Paris
法国, 2007

Svetlobna Gverilla
斯洛文尼亚, 2009

Lumina Light Festival
葡萄牙, 2013

Durham Light Festival
英国, 2013

Singapore Night Festival
新加坡, 2014

Photos ©: Dimitri Xenakis

漂浮之屋
the floating houses
奥地利 Austria

2013

乐巴尔托工作室为格拉茨2013年的施蒂里亚之秋现代艺术节规划并设计了接待场所，这片废弃的荒地曾经是旧海关总署的所在地。设计师为此规划了一条参观路线，并将园艺温室架空在3米的高度上，点缀着不同的步行空间。这些或透明或封闭的温室，可以容纳小团体的人们，使他们在光线的保护下，在此互相交流观点或吐露隐私。受到皮埃尔·肖代洛·德拉克洛的著名小说《危险关系》以及其中所表达的"必要的隐私需求"的启发，这些飘浮的温室呼应了此活动的主题"革命之后"，以及其脆弱而又必要的过程：重组、规范化、调适、商议并达成共识……

For Steirischer Herbst 2013, the contemporary art festival of Graz, the Atelier le Balto conceived the design for the welcome centre. Horticultural greenhouses, perched three meters high, form a parcourse, punctuating the different fallow spaces around the ancient customs house of the city. Opaque or transparent, the greenhouses welcome little groups of people who can exchange ideas or secrets under their luminous protection. Inspired by the famous Liaisons Dangereuses of Pierre Choderlos de Laclos, and the pressing need for intimacy, these greenhouses echo back to the over-arching theme of the exhibition, "after the revolution", and to its fragile required passages: reorientations, normalisations, adaptations but also consensus and negotiations...

Atelier le Balto
德国
Steirischer Herbst
奥地利

Photos ©: Winkelmeier

此基金会建造在一块面积30 000平方米的古老熔岩地块，位于兰萨罗特岛的塔伊切，最初是艺术家塞萨尔·曼里克在五个火山熔岩洞窟上搭建起的住所。基金会以受到1970年代风格影响的现代设计与当地民间建筑传统适切地融合在一起，透过配置有度的内院、花园和向着周围景观开放的视野与自然环境相互呼应，不断地进行对话。

Constructed over an old lava field of 30 000 m² ground, in Tahiche on the island of Lanzarote, this foundation was the artist César Manrique's house, which he erected on five volcanic bubbles. Harmonious synthesis of the tradition of local popular architecture and a modern conception, influenced by the 1970s, the building is in a permanent dialogue with nature by the interplay of the patios, gardens and openings with the surrounding landscape.

塞萨尔·曼里克基金会 /
césar manrique foundation - 1968

César Manrique (西班牙)

Photos ©: Sophie Barbaux

03

居民造景

INHABITANT LANDSCAPERS

居民造景
inhabitant landscapers

驻扎花园 /
the inhabited garden - 1979

Alain Bourbonnais (法国)
La Fabuloserie (法国)

Photos ©: **Jean-François Hamon**

建筑师阿兰·布尔博内为了置放个人的"非常规艺术"收藏品而创建了"奇幻园"原生艺术博物馆。在"驻扎花园"中,他在水景周围布置了弗朗索瓦·波特阿创作的"圆形徽章"、佩蒂特·皮埃尔的非凡"机械马戏团"、让·贝尔托勒设计的"风向标"、居勒·当卢的作品"小型非洲"和卡米尔·维达勒制作的真人大小的"亚当与夏娃"雕塑。

La Fabuloserie was dreamed up by the architect Alain Bourbonnais to house his collection of "non-standard" art. In "The Inhabited Garden", he arranged around a pool of water various works: the medaillions of François Portrat, the astonishing merry-go-round of Petit Pierre, the weather vanes of Jean Bertholle, the Little Africa of Jules Damloup, and the lifesize Adam and Eve of Camille Vidal.

pp.146-148

法国造型艺术家兼景观设计师贝尔纳·拉叙斯于1961年着手研究独立屋住宅及其景观整治。为此,他建立了一份环境调查清单,记录下一些由自学自创甚至特立独行的人所构思建造、堪称具有想象力与远见的景观环境。贝尔纳·拉许斯称这些人为"居民景观师",由此显现他对这些引人入胜的艺术作品的实际关注。这些居民景观师们回收利用废弃物、积累日常生活或天然材料(石头、贝壳、玻璃或瓷器碎片……),用以装饰美化个人家园,也为其增添趣味。

1974年法国人类学与人种学家克劳德·列维-斯特劳斯将"居民景观师"这个名称视为一项新的研究领域,它既非来自原生艺术,也不属于纪念性创作的范畴。这个称谓也对官方艺术与建筑的美学依据提出了质疑。这些雕塑和装饰元素来自创作者的想象力,显示出穷极思变的即兴创造能力,并诉说着个人的故事,民间艺术、建筑和宗教是最常见的灵感来源。

In 1961, the visual artist and landscaper Bernard Lassus begins a research on individual homes and their landscapes. In this context, he draws up an inventory of environments. One might characterize them as visionary, conceived as they are by autodidacts, even mavericks, as Lassus calls them, which shows his real interest for these spectacular works of art. Realized with salvage, recycling and accumulation of objects of everyday life or of nature (stones, shells, shards of glass or porcelain…), they enliven and decorate their residences.

The designation "inhabitant landscapers" was recognized in 1974 by Claude Lévi-Strauss as a new area of study distinct from monumental or outsider art. The newly identified art questions the aesthetic principles of art and official architecture. These sculptures and ornamentations, emerging from the imagination of their creators, show a spontaneous ability to work around a lack of means and to tell a personal story, often inspired by popular art, architecture and religion.

这股特殊的潮流并不仅限于西方,世界各地都有案例,其共通点在于使用从大自然中拾取的元素,以及将日常生活的废弃物加以转化而应用。

自20世纪末起,这些特殊艺术创作激发了知名艺术家的浓厚兴趣,他们随而创造出独具个人风格的景观艺术,通常展现于自家住所。无论是造型艺术家、导演或表演人士,他们和早期的居民景观师一般,支持并发扬"分享"的概念,乐于将作品展现出来,供人参观欣赏,同时也极为重视与他人的结识与交流。某些居民景观师改变了操作方式,不再仅限于运用回收材料,而是着手制作工艺或艺术品作为花园的装饰元素,并利用非凡奇异的植物造型,甚至运用一些特殊基地来自由挥洒独特而疯狂的创意。

This particular current is not a purely Western phenomenon. One finds it all over the world, with the common threads being the use of elements gleaned from nature as well as the transformation of refuse.

From the end of the 20th century, these art "outsiders" have aroused a great interest among recognized artists, who have given birth themselves to their own personal landscaping creation, often at their homes. Whether they are visual artists, directors or people of the theatre, they cultivate the notion of sharing, the desire that their realisations be seen and visited, while according a great importance to encounters and exchanges. They have certainly changed operational modes: not only utilizing recycled materials but also artisanally made decorative elements; growing vegetation in its most astonishing forms; and even dedicating themselves to a particular site to give free reign to their singular creative folly.

施华洛世奇水晶世界 /
swarovski crystal worlds - 1995
André Heller (奥地利)
Swarovski (奥地利)
Photos ©: Swarovski Kristallwelten & Anatol Jasiutyn

为庆祝施华洛世奇成立百年,这座令人叹为观止的水晶世界于1995年开幕,展示水晶各种巅峰造极的创意造型。水晶世界位于奥地利蒂罗尔州的瓦腾斯镇,而施华洛世奇首创人丹尼尔·施华洛世奇的第一个工作坊便是在该地建立的。这座水晶神殿占地4 000平方米,由当代著名的跨领域艺术家(亚历山大·麦昆、布莱恩·依诺、托德·布歇尔等人)改造的晶洞组成,入口丘陵上雕塑的巨人脸庞让参观者从一开始便坠入了梦幻世界……

This astonishing place dedicated to crystal in the most creative forms was inaugurated in 1995, for the centenary of the Swarovski company in Wattens, in the Austrian Tyrol, where the founder Daniel Swarovski created his first workshop. Composed of 4 000 m² of caves developed by great contemporary, multi-media artists (Alexander Mc Queen, Brian Eno, Tord Boontje...), this temple of crystal plunges the visitors, right from the entrance, into a dream world in the form of a giant face sculpted in the hillside.

p.149

查尔特勒修道院花园
the gardens of the chartreux
法国 France

1356-2015

这所建于1356年的查尔特勒修道院位于法国维尔诺夫-雷-阿维尼翁。查尔特勒修会是天主教的支派,严格奉行苦修,修道场所同时也是矿、植物并存的和谐空间。修道院中超过半数的面积都用于种植各种花草树木:果园、菜圃、药草园与游赏花园,一切都歌咏着造物主的神奇。

查尔特勒修士在修道院隐居,供修士独处的僻静小室面对着一个封闭式的小花园。这个拥抱自然的空间,不但是修士冥想灵修的所在,也是他们静心祈祷的园地。

Founded in 1356, the Chartreuse of Villeneuve Lez Avignon followed one of the most austere and rigorous orders of Catholicism. This Carthusian monastery was also an equilibrium between the vegetable and mineral world: more than half of its surface was planted. Orchards, vegetable gardens, herb gardens and pleasure gardens, all were an homage rendered to the Creator.

The monks who lived cloistered in the solitude of their cells had at their disposition a little enclosed garden onto which their rooms opened. This opening into nature was both a space of meditation and an aid for prayer.

Chartreuse de Villeneuve Lez Avignon
法国

Photos ©: Catherine Gilly-Corre p.150
Alex Nollet p.151

时至今日，过去的修道院已经变成了"国立表演艺术著作中心"，不再提供修士隐居，却转而成为接待剧作家和各类艺术家居留的地方。在这个交会邂逅的场所，花园一直都占有非常重要的地位。有些住在从前修士小室的艺术家，依然维续着这个私秘花园的传统，迪迪埃乔治·伽比里即为一例，他于1993年让人在花园里种植了红玫瑰。

Today the site, a national centre for play-writing, no longer accommodates monks but offers lodging to playwrights and artists of different disciplines. In this place of crossroads and encounters, the gardens have always played a very big part. Some residents, housed in the old cells, perpetuate the tradition of intimate garden, like Didier-Georges Gabily, the well known writer, who had red rosebushes planted in 1993.

乐土
paradise
法国 France

1955

雷欧珀德·图尔克是法国卢贝隆山区中阿维尼翁－卡布雷尔村庄的农夫，有个儿子是餐饮业者。从1955年起，他就开始捡拾儿子丢弃的扇贝壳，并结合碎瓦片拼接在一起，以此装饰自己的田庄。他当时是为了避免浪费，如今的说法则是资源回收。1956年北半球的寒冬创低温记录，普罗旺斯地区有许多橄榄树因而冻死，雷欧珀德·图尔克的土地也难逃劫数。他因此做了一个决定：不再种植橄榄树，而是在整个庄园实现他个人的"乐土"计划。

In 1955, Léopold Truc, farmer in Cabrières d'Avignon, in the Luberon area of southern France, begins to decorate a stone shepherd's hut with mosaics using scallop shells thrown out by his restaurant-owner son, and pieces of broken tiles to prevent waste: today, we would call it recycling. The winter of 1956, marked by record cold in the northern hemisphere, causes the loss of many olive trees in Provence, including ones on his property. He decides not to replant but to develop his personal Paradise project on the whole site.

Léopold Truc
法国

Photos ©: Sophie Barbaux

从入口便可见到取名"乐土"的庄园标示牌。循着一条林荫小径，便可通往被雷欧珀德·图尔克唤作"小棚屋"的房子，其他六条小路与林荫小径垂直相交，形成了庄园空间的组织结构，每条路面上都铺满了贝壳和石头。许多陶盆堆叠成柱，呈直线排列，仿佛取代了树木，还有不少造像雕塑，这儿一座，那儿一座的，其中尤其值得一提的是马槽，以及向法国戴高乐将军致敬的洛林十字架。随着时光推移，一些小型建筑物逐渐完成：冬天的避寒处及阳台、大小如岗哨的小教堂、他自己的纪念碑……还有个小观景楼，雷欧珀德·图尔克从那里可以观赏自己的作品，眺望周围的乡间景致。

At the entry, a sign gives the name "Paradise". A little path leads to his cabin, as Léopold Truc calls it. It is paved with shells and stones, like the six other paths that run perpendicularly and structure the space. Numerous rows of columns made of piled-up terra cotta pots, seem to have replaced the trees. Here and there Truc has placed figurative sculptures, notably a creche and a cross of Lorraine, a symbol of General de Gaulle's Free French forces. Small structures come to light little by little: a shelter for winter, and its terrace, a chapel as big as a workman's hut, his own memorials… And a look-out tower from where Léopol Truc can contemplate his work and the surrounding countryside.

这座矿物花园由紫、灰、赭等色调组成，是一位崇尚"生存经济"的艺术家的园地。他同时也是幻想国度的建造者，以梦想塑造出他心目中的人间乐土。

This mineral garden, with purple, teal-grey and ochre tones, belongs to an artist gifted with the ability to make use out of others' waste and also of a builder of the imaginary, who has dreamed and fashioned his earthly paradise in his own image.

居民造景 / 执念

居民造景 / 执念

岩石花园
rock garden
印度 India

1957

"岩石花园"这个疯狂的故事起始于1957年，它就离昌迪加尔这个由柯布西耶从1951年起负责规划建造的现代新城几千米远。尼克·昌德是城市公共交通机构的员工，每天都骑着单车前往昌迪加尔。他沿途搜集了许多石头，全都堆放在一块禁建地里，除此之外，由于建造新城需要空间，许多村庄因此人去楼空，丢弃了不少物件，他也在垃圾堆积场和村庄废墟里拾取废弃物，一并和石头汇聚在一起。

每天晚上，在燃烧轮胎的火光照耀下，他开始秘密进行为童年圆梦的工作，以水泥为底，盘子、杯子或石头等等做装饰，创造出第一批巨大的人物与动物塑像。

In 1957, the mad adventure of the Rock Garden begins a few kilometres from Chandigarh, the modern village created in 1951 by Le Corbusier. Nek Chand Saini, modest public transportation employee, goes there every day on his bike. Along the way, he collects rocks that he begins to amass on an unbuildable piece of property, as well as trash recuperated from dumps and abandoned villages, that made room for the new city.

Each night, realising his childhood dream by the light of burning tires and in total secrecy, he gives birth to his first big statues of men and animals, in cement adorned with pieces of dishware, glass or pebbles...

Nek Chand Saini
印度
Nek Chand Foundation
印度

Photos ©: Raw Vision magazine

随着时间过去，其创作内容越发丰富多样，数百件新作品陆续问世，真人真物大小的塑像矗立在一系列的庭院里，唤起人们对古老印度传统的记忆。

1975年，昌迪加尔市政府决定清理整顿整个基地，从而发现这位自学自成的艺术家不同凡响的艺术创作计划。场所的神奇之力发挥了效果，尼克·昌德因此获得了新城总建筑师的支持，不仅允许他留在原地，同时还付给他薪水，提供基金与协助，让尼克·昌德能继续在16公顷地的空间里继续创作发展这个令人惊奇的雕塑花园。今天，这个花园已成为印度第二大观光景点，仅次于泰姬陵。

Over the years, the project prospers and hundreds of new, lifesize works populates a succession of courtyards, evoking ancient Indian traditions.

In 1975, the city of Chandigarh decideed to clean the site and discovered the amplitude of this exceptional artistic project, created by an autodidact. The magic of the site works, and Nek Chand Saini receives the support of the head architect of the city, who, not only allows him to remain there, but also obtains a salary for him, as well as the funds and help to pursue the development of his astonishing sculpture garden, on 16 hectares. Today, it is the second most visited site in India, after the Taj Mahal.

罗伯·塔坦花园博物馆
the garden museum of robert tatin
法国 France

1963

罗伯·塔坦在家乡法国马耶讷省的博卡日田园购置一块乡村地时已经60岁了。这位求知欲极强的艺术家曾经受过多种训练，他在这块土地上构思创造出绝无仅有的作品，与自然环境和谐融成一体的同时，也向各种不同的文明与文化致敬。他曾周游世界各地，种种经历不仅为这项以人文关怀为本的计划带来素材，也赋予它世界共通的语言。罗伯·塔坦选择了以上色钢筋混凝土作为材料，来发展这个世界语言，因为这个材料能为创作提供极大的自由度。

在这个花园里，一切都具有象征意义。花园入口便是"伟人之道"，路径两旁矗立着从维钦托利、毕加索到儒勒·凡尔纳笔下的英雄人物等各种塑像。

Robert Tatin is 60 years old when he acquires a little piece of property in Cossé-le-Vivien, in the bocage of Mayenne, his birthplace. This artist with an insatiable appetite and a varied education undertakes a unique work that renders homage to different civilizations, while at the same time inserting itself harmoniously in nature. His numerous voyages all over the world give him both the subject matter of his humanist project as well as the universal language that he develops, choosing as prime material painted reinforced cement, which offers a great liberty of creation.

In this garden, everything is symbol. One enters by the path of the giants, from Vercingetorix, the Celtic hero, to Jules Verne, to Pablo Picasso.

Robert Tatin
法国
Musée Robert Tatin
法国

Photos ©: Musée Robert Tatin

走在"伟人之道"上,远远望去,映入眼帘的是园内的博物馆中心,还有"冥想园"。"冥想园"周围环绕着浮雕装饰的墙面,诉说着东西方的神话故事。中央呈十字架形的大水池外围以地球自转的方向安置了一些小型雕塑,对应着一年内的每个月份,令人怀想起大自然鬼斧神工的运作机制。"世界的圣母"雕像在此高耸入云,天地宇宙彷佛触手可及。

罗伯·塔坦设计的"石园"和"蝴蝶栖所"带领观者深入这段奇妙的旅程,在大自然伴人左右的情境中,引人省思人类在过去与现在的景况,进而思考未来的方向。

Next, a perspective opens upon the heart of the museum, with the Garden of Meditations, enclosed by walls decorated with bas-reliefs that evoke the myths and legends of the East and West. There is also a large central basin, in the form of a cross, around which one walks in the direction of the rotation of the earth, viewing small sculptures that recall the work and the natural events belonging to each month of the year. Notre-Dame of the Whole World rises towards the sky, bringing the cosmos close at hand.

The Rock Garden and the Butterfly Refuge complete the itinerary conceived by Robert Tatin to plunge us into a reflection on the past, the present and the future of humanity, without losing nature as our companion.

170 inhabitant landscapers / PASSION

居民造景 / 热忱 171

荷花园
lotusland 美国 USA

1941

这个位于蒙泰西多的庄园在19世纪末时属于某位园艺家，他在此地种植了特别显眼的高大棕榈树，1920年代之际，意大利古典式花园是这个基地整治的灵感来源。1941年波兰歌剧名伶甘娜·瓦拉斯卡买下这个庄园，之后长达43年的时间里，她在这个占地1.5公顷的土地上创造了许多花园，直到1984年去世为止，并将此庄园重新命名为"荷花园"。

At the end of the 19th century, this estate, situated in Montecito, belongs to a horticulturist who plants in particular immense palm trees. During the 1920s his project is inspired by classic Italian gardens. The Polish opera singer and society woman, Ganna Walska, bought the property in 1941 and for 43 years until her death in 1984, she creates numerous gardens on these 1.5 hectares, renaming it Lotusland.

Ganna Walska
美国
Lotusland
美国

Photos ©: Lotusland

直至今日,荷花园环游世界十八地的植物之旅仍继续邀人探访寻幽。一系列异国植物接连着亚热带植物(凤梨科、芦荟、仙人掌、大戟属、蕨类植物等等),以及柑橘园、日式花园、摩尔式露台……有些空间展现出怪异、甚而戏剧性的景致:有个花园里,动物塑形的植物围绕着一个标示了黄道十二宫图案的真实时钟;在绿地露天剧场里,矗立着一座座造型怪诞的石雕;还有岩石、珊瑚错落有致的蓝色水泊,边缘以鲍鱼壳绕圈装饰,奇形怪状的巨大贝壳鹤立其间……

With 18 stops around the world, Lotusland invites the visitor on a voyage. Collections of exotic and subtropical plants (Bromeliacae, Aloe, cactus, euphorbias, ferns, etc.) succeed one another in series. A citrus grove placed side by side with a Japanese garden and a Moorish terrace. Certain spaces are arranged off centre, often in a dramatic way: a flowerbed of animal topiaries frame a real clock that displays the signs of the Zodiac, a theatre of greenery where grotesque statuary propagates, or again a blue lagoon of rocks and coral where big, strangely sculptural shells overhang a border of elms…

176 inhabitant landscapers / COLLECTION

陶花园
ceramic flower garden
法国 France

1983

这个雅致迷人的花园位于尼斯附近乡间的加蒂埃村庄，在占地1 500平方米的缓坡地块上逐渐展开，并由12株百年橄榄树勾勒出主要的空间结构。"陶花小径"展现出植物与矿物和谐结合的基调，引领人们走进花园的天地。一系列的小园圃接续交替，将近800种稀有或野生植物在此蓬勃生长，伴随着自然生物造型的涂色黏土雕塑，也点缀着一些随着时间而捡拾或搜集而来的化石、贝壳、珊瑚、镜子和各种玻璃珠、玻璃球。

Located in the countryside behind Nice, in Gattières, this gracious garden, structured by a grove of centuries-old olive trees, unfolds across 1 500 m² of gently sloping terrain. There, one penetrates an alley of pottery flowers, which sets the tone to this successful marriage of vegetable and mineral. In a succession of little gardens, a collection of about 800 rare and wild plants blossom, accompanied by sculptures of coloured clay, their forms inspired by nature. Mixed in are fossils, shells, corals, mirrors, marbles and balls of all sorts, gathered over the years.

Anne-Marie & Raymond Deloire
法国

Photos ©: Anne-Marie Deloire

参观这个花园的旅程充满了惊喜,陶土化身为抽象或具象的嫩芽与花冠,以及经常在花园里穿梭来去的动物。这些动植物无惧风霜雨雪,四季常在,是安-玛丽·德罗荷丰富的想象力所创造的成果,诸多细节供人探索玩味,令人流连忘返……

The visit is an unexpected voyage, the earth metamorphosing in abstract or figurative shoots and corollas, and in animals habitually populating the gardens. This fauna and flora, evergreen in all seasons, was conceived and created by Anne-Marie Deloire, who offers to the visitors a multitude of details to discover, making one want to linger.

仙人掌花园
the cactus garden
西班牙 Spain

兰萨罗特岛是塞萨尔·曼里克的家乡，这里满布着这位同时是画家、雕塑家与建筑师的印记。塞萨尔·曼里克曾付出大量心力来保护自己家乡的文化特质，并致力维护这个加那利群岛中的岛屿自然景观。他所设计的建筑与作品在岛上交错林立，巧妙地融入当地环境，而"仙人掌花园"则是最后完成的创作。

入口的迷宫阵凸显出这座花园的与众不同，并放缓了游客探访花园的脚步。过了迷宫，花园全景豁然出现眼前，圆形剧场式与旧火山岩采石场的地形完美结合，显示岛屿火山的地势特质。

Lanzarote was profoundly changed by the presence of César Manrique, painter, sculptor, architect and local boy determined to work for the conservation of the cultural identity and the preservation of the landscape of this Canary Island. He punctuated it with structures judiciously integrated into the site and also with works of which "The Cactus Garden" is the last.

Right from the entrance, a maze marks the garden's singularity, slowing one's discovery of the rest of the ensemble. After the maze, the garden reveals itself to be a circular amphitheatre that follows the form of this ancient quarry of volcanic cinder while evoking insular volcanoes.

César Manrique
西班牙

Photos ©: Sophie Barbaux

每个石砌梯层既是散步步道，也是观赏仙人掌和其他当地多肉植物的所在，仙人掌系列共有6 200棵，囊括了800种不同品种，令人叹为观止。

然而，此基地最迷人的是石柱场景，以红砾石堆积、压实而成的红石柱，矗立在花园中央。巧妙而协调的植物搭配与布置也毫不逊色，如雕塑般的形体展现在游客眼前，魅力十足。花园里还有一个研磨烤谷物面粉的旧时风车磨坊，立在碎石堆砌的围墙上，而一些环礁湖则使这个戏剧性十足的植物交响曲更为完美动人。

Each row of stone seats is a place both to stroll and to view an exhibition of a surprising collection of 6 200 cactus, featuring 800 species, and other types of native succulents.

But the most fascinating part of the site is the scenography of the monoliths of compacted red gravel that stand in its centre. The "staging" of plants, by the harmony of their associations, offers sculptural forms to the visitors. An old windmill for grinding gofio, a Canary Island cereal, crowns the stone wall, and small lagoons complete this botanical theatre.

展望小屋
prospect cottage
英国 United Kingdom

这栋小屋位于英国南部邓杰内斯,沥青黑面木板配搭金丝雀黄的门条窗框,让德里克·贾曼在一个近乎偶然的机缘里买下了它。这位执导多部地下电影的导演是英国后朋克年代的代表人物,而他同时也是位画家、作家和剧场工作者。身染艾滋病后,他在这栋小屋里度过了人生最后阶段中的美好时光。

Purchased almost by chance by Derek Jarman, this tarred-clapboard cottage with canary yellow window frames, is situated in Dungeness, in southern England. The underground film director, icon of the British post-punk generation, but also painter, writer and man of the theatre, diagnosed with AIDS, would pass the best moments of the end of his life there.

Derek Jarman
英国

Photos ©: John Siddique

小屋的所在地非比寻常，距离海岸和核能厂都仅几百米之遥。四周景物乏善可陈，只有几户捕鱼人家的房舍，放眼尽是沙土砾地，仅仅蹿出几株枯瘦的植物，受寒风烈日蹂躏。

The place is a little banal, a few feet from the sea, but also from a nuclear power plant. The landscape is desolated, the cottages of fishermen surrounded by gravel where only a few skimpy plants grow, beaten by the winds and wilted by a burning sun.

自从德里克·贾曼捡拾了第一块燧石、捞起了邻近沙岸的一块漂流木之后,这个景观庭园便从此诞生了。卵石、水手用的钉耙、厚木板、生锈的工具或海浪磨平的木桩构成了第一批景观元素。这些海边寻获的宝藏被用来支撑苗木,或成为桩柱、围栏、雕塑品,——竖立在这个空旷的基地上。紧接而来的工作是栽种新植物的实验,这些植物必须能在严酷的气候下生存,以便搭配当地原生植物如"荒野之王"海甘蓝,以及银香菊、桂竹香、山楂、黑刺李树。

The idea for a garden began with the finding of a first arrowhead and of some driftwood on the nearby shore. Thus, pebbles, bargeman's hooks, beams, rusted tools or posts smoothed by the waves are the first landscaping elements. These marine treasures become stakes, posts and trellises, sculptures that rise up in this site where nothing stops the gaze. Next comes the experimentations with new plants. If they can tolerate the climate, they will flank natives such as sea kale, the king of the marsh, but also santolinas, gillyflower, hawthorn and blackthorn.

这栋小屋也是德里克·贾曼抒情表意的所在，墙上粘贴了约翰·邓恩的诗歌《太阳升起》，和此地自然环境相互应和，为这位艺术家提供了宁静的避风港，让他得以分享内心的世界。

The cottage is itself also a place of expression, welcoming on its walls the John Donne poem, "The Sun Rising", echo of nature, which offers itself to the artist as a haven of peace and shared intimacy.

居民造景 / 转化 201

石园
rock garden 法国 France

2002

最初，这里是一块荒地，遍布树龄较短的松树，以及用作基地堆肥的常春藤。在一次为防火而实施的灌木、杂草清理后，一个古老的凿切石采石场赫然现身，工作人员随后进行挖掘工程，试图寻找可能连接邻近采石场的地道遗址，这个基地从此呈现顺着岩脉倾斜的坡地外观。

Originally, a fallow field overgrown with young pines and ivy served as a compost pit. After a controlled fire to clear the area, a former stone quarry was uncovered. Digging operations to find tunnels that might have linked it to nearby extraction sites left the quarry looking like an inclined plane that followed the direction of the vein of the rock.

Alpilles
法国

Photos ©: DR

随着时间推移，不同的景观元素逐渐在这个矿质空间定位：日式灯笼、立石（例如支石墓）、卵石河床、休憩亭、岩石上的埃及式石雕门，引领人进入不可见的世界……一个瀑布、一个小池塘（池塘里纸莎草、睡莲、荷花遍生，并有金鱼穿梭其间）和一些当地植物，都恰如其分地为这个奇特的石园增添景致，邀请人们神游至遥远的国度，在那里，时光仿佛静止于某个古老时刻。

Over time, different landscaping elements came to inhabit this mineral space: Japanese lanterns, stones raised like dolmens, a pebble river, a pavilion for rest and relaxation, an Egyptian door sculpted in the boulders leading to the invisible world... A waterfall, a little pond populated with papyrus reeds, lilies, lotuses and koi, the surrounding area judiciously planted with local species complete the singularity of this rock garden, a place evoking distant lands, where time seems to have stopped.

埃吉莱花园
the garden of éguilles
法国 France

当马克斯·梭兹最初来到此处时，这个花园的基地还只是个废弃荒芜的菜园，只有几条水泥制的灌溉小渠道、几株果树、一棵棕榈树和几棵大树，偶尔看得到羊群前来吃草。随着岁月流逝，这位艺术家的花园逐渐蜕变，转化为诗意创作的空间，植被繁茂而郁郁葱葱，包括野生的和经过培育的植物，而百余种既似物品又似雕塑的新"居民"也在当中落地生根。

On the arrival of Max Sauze, this site was an abandoned vegetable garden, with little cement irrigation canals, fruit trees, a palm and a few big trees, where sheep sometimes came to graze. Over the years, this artist's garden slowly metamorphoses, becoming a poetic creation with lush vegetation, both wild and domesticated, among which a hundred new inhabitants have taken root, half-object, half-sculpture.

Max Sauze
法国

Photos ©: Max Sauze & Anne Le Berre

这些作品都是应用回收材料进行创作的成果，其材质繁多：纸张、蜗牛壳、金属、小石块、松果，甚至硬币也包含在内。各类作品经过精心设计，乱中有序，或置于树干之间、挂在树枝上，或和墙壁与围篱融为一体，创造出前所未见的新叶、果实，和令人耳目一新的色彩。随意穿梭在步道间，每逢迂回转弯处便可看见"蜗牛捕捉器"、"椴树池"、"文字千层派"、"圣伊奥尔的堆叠"、"水盆"……每件作品都仿佛一个亲密的内在世界，诉说着美妙动人的故事。

The works are all conceived with salvaged material, as varied as paper, snails, metal, pebbles, pinecones, or even coins. In a rigorously organized disorder, they nestle between trunks, hang off branches, fade into the walls and the fences and invent new leafing, unknown fruits, surprising colours. In the course of wandering, these works, with names such as "the snail swatters", "the lime-tree pond", "the writing leaves", "the piles of Saint-Yorre"... like "the bascine", create their own intimate universe, filled with the marvels of the story they tell.

促使大自然与人为艺术品共生共存是马克斯·梭兹所冀望并努力达到的成果，这样的关系随着岁月的脚步持续发展着，或消失、修整、或彻底改头换面，或在偶然间诞生新貌。这座藏身于埃吉莱村庄里的静谧花园虽然外貌日益变化，精神却永驻不朽。

This cohabitation, desired and cultivated, between nature and artifice, continues from one day to the next with losses, repairs, radical transformations and births due to chance. All these events modify the site, without changing the spirit of the garden of silence hidden at the heart of the village of Éguilles.

212 inhabitant landscapers / TRANSFORMATION

居民造景 / 转化 215

黄磨坊花园
the garden of the moulin jaune
法国 France

2003

这个坐落于克雷西-拉-沙佩勒的奇特花园占地超过3公顷，沿着塞纳-马恩省的大莫兰河而延伸。花园的调性从入口处两尊小丑雕像的巨大尺寸便一锤定音，它们是出自园主斯拉法·帕拉尼某项表演中的人物，幽灵般地簇拥着入口宏伟的木门。

散步步道逆流而上，一个个不同主题的小花圃依序点缀着河岸边的风景。这里弥漫着近似18世纪末的氛围，景观空间的组织随着一些令人惊奇的点缀性建筑而展开，撩起人们对远方国度之旅的怀想。

This strange garden is situated in Crécy-la-Chapelle, on around three hectares along the Grand Morin, a river in the department of the Seine-et-Marne. From the entrance, the tone is established by the outlandishness of the two sculptures of ghostly clowns, left over from a theatrical piece of Slava Polunin, the master creator of the place. The sculptures frame an imposing wood portal.

A promenade goes upstream, accompanied by a succession of small thematic gardens that punctuate the river banks. With an ambience reminiscent of the end of the 18th century, the landscape includes surprising fabriques, also evocative of distant voyages.

Slava Polunin
俄罗斯

Photos ©: Anna Hannikainen, Vladimir Mishukov, Jurijs Suhodolskijs, Suvorova Maria

蜿蜒于水上的甲板形成一条500米长的红线，在不同转弯处接连出现着各种场景：印度拉贾斯坦邦白色神庙、吉普赛人古老的酒红色篷车、月形的薰衣草园、阴森森的动物造型的观景台，和一座佛教僧侣赠献的华丽韩国寺庙。每个场所都各有专属的植被栽种和特殊颜色，大自然的缤纷色彩尽现眼前。

A sinuous area, 500 metres long, provides the unifying element of a diverse collection: a gleaming white temple of Rajasthan, an antique wine-red gypsy caravan, a lavender moon, a look-out point with dark animal forms, and a flamboyant Korean temple donated by Buddhist monks. Each stage has its own vegetation, each with a specific colour, using the entire palette available in nature.

这座花园既是观赏演出的场所，也是让人休憩和冥想的空间，透过设计师、造型艺术家、建筑师和演员们的默契付出，和谐创造而成，却又生生不息、变化万千。苏联戏剧大师尼古拉·埃夫尼诺夫的艺术人生是此花园创作的灵感来源，这个具有分享精神的港湾也是无常的国度，不仅是实验性的创作空间，更成为交谊与游乐的场所。在这里，发挥想象力与实现梦想便是至上规则，并且永无止境……

A place for performances, for rest, even for meditation, these gardens, conceived with the help of designers, mixed media artists, architects and actors, are in a continual state of becoming. Inspired by Nikolai Evreinov, a Russian director and producer who made his life a work of art, this shared haven is the realm of impermanence. It is at once a place of experimentation, of conviviality and of play where imagining and producing one's dreams is the rule, without end.

位于巴黎的拉维列特屠宰场于1974年关闭，1987年原基地出现了一个面积33公顷的同名公园。建筑师伯纳德·屈米从三个网格系统出发为公园进行构思设计，其中一个网格决定了26座鲜红色馥丽小筑的建造位置，在这个巴黎最大的绿色空间里，和各种浓淡不一的绿色调形成美妙的对比。

On the former site of the slaughterhouse of La Villette in Paris, closed in 1974, an eponymous park of 33 hectares is born thirteen years later. The architect Bernard Tschumi conceived it through a triple system of frameworks, one of which determines the placement of 26 follies, bright red, contrasting joyously with the verdant plaid of the vegetation forming Paris's biggest green space.

馥丽小筑狂想曲 / follies - 1987
Bernard Tschumi Architectes (法国&瑞士)
Parc de La Villette (法国)
Photos ©: Sophie Barbaux

04
馥丽小筑
FOLLIES

馥丽小筑
follies

花园凉亭 / garden pavilion - 2003
Paul Raff Studio (加拿大)
Artists' Garden Cooperative (加拿大)
Photos ©: Steven Evans & Brita Ralph

这个位于多伦多的当代小筑，其设计灵感来自东方哲学。它隐没在一座私人花园内，仿佛避风港般宁静而适合沉思默想，却也可以成为艺术表演的舞台。此木格结构体以未经任何处理的雪松搭架而成，创造出流光与疏影交错的曼妙效果，而其建造方位的选定有利于避开夏季直射的艳阳，却能捕捉冬季斜照的暖阳。凉亭的多处开口让人和大自然之间产生一种亲密关系，而镜面平桌映照着苍穹，使人垂下眼帘便得以端视天空……

Nestled in a private garden in Toronto, this contemporary fabrique is inspired by oriental philosophy, offering calm and contemplation but also the possibility of becoming an artistic scene. The structure of the haven in untreated cedar trellises plays with shadow and light, oriented to diffract the rays of sunlight in summer and to capture the warm and low light of winter. The multiple openings create an intimate relationship with nature, accentuated by the table-mirror, which reflects the zenith, permitting viewers to lower their eyes towards the sky... and woven into a living sculpture.

从景观层面而言，馥丽小筑（folies）最初指的是坐落在花园里的小型豪华休闲住宅，作为短期居所之用，必须具备一定的私密性。人们刻意地使馥丽小筑的建筑独具创意，甚而怪诞夸张。这个具有暗示性的名称来自拉丁文foleia（枝叶搭建的小屋或树荫庇护的场所），以此称呼位于茂密叶丛下的房子，而这个用词的首度使用则可追溯到16世纪。

在18世纪，许多私人园林以"英式"或"英中式"风格而建成，园内常见的装饰性建筑也被冠以"馥丽小筑"这个名称。其中某些依然保留居住用途，不过大多数都更加缩小，局部提供家居功能，在大多数情况下主要作为娱乐休闲所用。这种建筑物引人注目，造型或仿效古代以来西方世界和异国他方的建筑，或以哲学、文学和宗教为主题而建造。大自然的丰富形态同样也是设计灵感的来源。

Originally follies are, in the landscaping context, small and luxurious pleasure buildings situated in a garden, meant for temporary visits when intimacy is required. Their architecture is intentionally original, even extravagant. The evocative name seems to come form the latin *foleia* (leaf or leafed), describing a house in the foliage, and its first use dates from the 16th century.

In the 18th century, this term is also applied to garden fabriques that are beginning to be seen in private parks developed "in the English style" or "Anglo-Chinese". They certainly still have the function of habitat, but are mostly smaller; additionally, while they can be domestic, they are more often intended for pleasure. Attention-drawing structures, their forms are borrowed from Western architecture back to antiquity and also from exotic countries, or they are based around philosophical, literary or religious themes. The vocabulary of nature is also a rich source of inspiration.

18世纪末的出版品中呈现出多种类型的馥丽小筑（洞穴、古神庙、中式屋宇、露天剧场、金字塔、废墟……），为各项试验与诠释应用开启了大门。事实上，这些小型建构物成为景观组织的元素，也影响了植物的配置方式。根据所在位置的精心安排，它们可以成为游艺区的景观焦点，也可以在一条小径、一座小丘、一个树丛的拐弯处创造惊艳效果，为散步路径巧妙穿插一系列富于变化的氛围。馥丽小筑不仅成为优美如画的景观视点，让人以不同的方式感受花园，同时也是漫步者的休憩点、天候不佳时遮风避雨的场所。除了供人观赏沉思和休闲娱乐的用途之外，馥丽小筑也可通过其位置的设定来形成一条能够引导思考的散步路径，将启蒙时代所重视的主题纳入其中。

这股风潮逐渐趋于没落，自20世纪末期起却再度复苏，这种小型建筑重新驻入公园和私人花园里，起初先在城市空间占据一席之地，继而甚至尝试不同的尺度，现身于大型景观或区域景观。馥丽小筑的功能经常仅限于组构空间、美化装饰和休闲娱乐等用途，但有时也可能产生其他作用，如反映当代文化与社会关注课题：它们向当下世界的荒谬、粗暴以及生物多样性的现况提出质询，亦或成为供人社交往来、会晤、用餐的场所，甚至变成别出心裁的度假地点。馥丽小筑与当今的建筑和艺术创作一样，在造型和材质上具有多样变化，并再度成为短暂或长期实验的资源。

At the end of the 18th century, publications classify different types of follies (grottos, ancient temples, Chinese pagodas, green theatres, pyramids, ruins…), opening the door to experimentation and interpretation. In fact, these little constructions participate in the landscaping composition just like the placement of plants. Their position in the garden choreographs poles of attraction, surprises at the turn of a path, hill, or grove, punctuating the promenade with a series of ambiances and atmospheres. Follies offer picturesque points of view for understand the garden differently, as well as places of rest, shelter from bad weather. In their role as a medium for contemplation and for leisure, their placement can also delineate a path towards reflection, inspired by the themes dear to the Age of Enlightenment.

Going out of style progressively, follies have come back since the end of the 20th century to populate both public and private parks and gardens, all the while taking a place in the urban space. Designers even go so far as to use follies to experiment with different scales, such as that of the large landscape or of the territory. Their function often remains uniquely as an organisation of the space, as a means of beautifying or simply as a pleasure, but they can also have other purposes, such as reflections of contemporary social and cultural preoccupations. They can be used as a means to question the absurdity and the brutality of our world, to highlight its biodiversity. Follies can be places for conviviality, encounter, or snacking, or even an original vacation spot. Their forms and their materials are as varied as current architectural and artistic creation. Ephemeral or perennial, follies are new sources of experimentation.

视听室a / auditorium a - 2014
Claire Dehove & Daniel Deshays, Loraine Djidi, Cléo Laigret (法国)
Direction régionale de l'Équipement & DRAC Languedoc-Roussillon (法国)
Photos ©: Wos / Agencedeshypothèses

就大地景观的尺度而言，"视听室A"犹如一座共鸣小筑，将位于塞拉的高速公路立交桥转化为艺术品，供人欣赏、聆听……这个供人聆听的空间位于拉扎克石灰岩地丛生的柏木和橄榄树之间，它收纳周遭的声音，并透过电声装置与小室本身制造的声响相融合，再播放出混音创作。它同时也是一座观景台，因此能够引发多重感官效果，不断地更新、创造出听觉与视觉交织的美妙乐章。

Working on the scale of the large landscape, Auditorium A is a fabrique of resonance that transforms a highway interchange near the town of Ceyras into a work of art to look at and listen to. In the middle of the cypresses and olive trees of the limestone plateau of the Larzac, this "sound studio" samples and amplifies ambient sounds, mixing them with sounds that it produces, diffusing and rediffusing the mix through an electric-acoustic apparatus. A multi-sensory lookout point, it offers a sonorous and visual score, ceaselessly renewing itself.

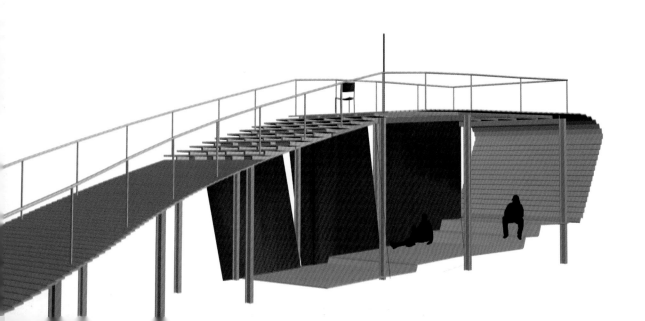

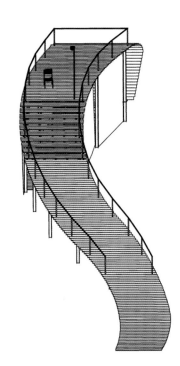

门外天地
outlandia
英国 United Kingdom

2010

这个奇特的结构体现身于苏格兰高地的本尼维斯山（不列颠群岛的最高峰）里面的一个混合着挪威杉和落叶松的森林里。一条穿越森林的小径将人引领到一座木桥上，而木桥的尽头设置了一扇关闭的门。

A strange structure has been installed in this forest of Norwegian spruce and larch, situated in the Scottish Highlands at the foot of Ben Nevis, the highest summit of the British Isles. A boardwalk that leads through the forest to a bridge ends in a closed door.

London Fieldworks & Malcolm Fraser Architects
英国
Glen Nevis
英国

Photos ©: Niall Jacobson

这个门界的设计旨在促进人类与大自然之间自发地建立起关系，并让这份联系在苏格兰林地里自由发展，仿佛门本身就足够开启一切。此作品除了在形体上受到传统日式建筑的启发，其创作灵感也来自孩童之间互相传说的故事，让人投入门外天地的想象旅程……

Meant to facilitate a spontaneous relationship with nature, developed in Scottish forestry, this threshold, as if sufficient in itself, is inspired in its form by traditional Japanese architecture, and also by stories children tell, eager to engage in imaginary adventures, from the other side…

蛇形树屋
tree snake houses
葡萄牙 Portugal

2013

位于葡萄牙北部的佩德拉斯萨尔加达斯公园是一个古老的温泉乡，在一个生态建筑项目的背景下，利用板岩和木头等当地材料重新整治建设后，展现了令人惊奇的新貌。十二座当代小木屋隐身在树林里，巧妙地与基地地势结合。其中两座悬置在空中，人们漫步森林之际，便赫然进入视野，令人眼前一亮。木屋的外形摒弃传统的垂直正交结构，而是采用蛇状造型，以便自然地溜爬进树干之间，在攀升之际挑战重力的作用。

In the Pedras Salgadas Park, in the north of Portugal, a former hot springs has undergone an astonishing renewal. In the context of a eco-construction project, it utilizes local materials such as slate and wood. Twelve contemporary pavilions hide in the woods and, playing with the topography of the site. Two of them, in particular, surprise us, hanging into the visual field of the passer-by. Their forms have abandoned classic orthogonality to slip naturally between tree trunks, like snakes, defying gravity.

Luis & Tiago Rebelo de Andrade
葡萄牙
Pedras Salgadas spa & Nature Park
葡萄牙

Photos ©: Pedras Salgadas Spa & Nature Park

2012-2013

生物拟态
biomimicry
英国 United Kingdom

受大自然与几何学所启发，安东尼·吉本在森林环境中设计出能提供居住的生态建筑。每间树屋环绕着一棵树木发展，透过生物拟态的方式融入森林景观中。

在"栖息"小屋中，一个螺旋状的阶梯沿着树干爬升而上，抵达360度的观景台，使人视野得以在树林的顶层徜徉。以雪松木片建构的"胚胎"树屋则采用大颗种子的造型，仿佛紧紧依附着树干的木瘤，这个遮风避雨的温馨小室让人们重新找回和大自然之间的关系。

Inspired by nature and geometry, Antony Gibbon imagines ecological and inhabitable structures integrated into a forest environment. Each work adopts a tree around which it develops, thus melting into the landscape through biomimicry.

For example, "Roost" features a spiral staircase that gives access to a swooping panorama that one can enjoy like a bird perched over the tree canopy. And "Embryo" adopts the form of a fat seed made of cedar shingles, hangs like a wood knot from a tree trunk. Protective cocoon, "Embryo" allows one to rediscover a forgotten relationship with nature.

Antony Gibbon, antonygibbondesigns
英国

栖息 / Roost
p.240 & p.241 上/top

胚胎 / Embryo
p.241 下/bottom

Images ©: Antonygibbondesigns

超级国度&自发城市
super kingdom & spontaneous city
英国 United Kingdom

布鲁斯·吉尔克里斯特和乔·乔尔森组成"伦敦原野行动"设计团队,积极发展一些将大自然与社会环境结合在一起的计划。2008年,他们以"超级国度"这个作品来揭发独裁者的奢华宫殿,这些专为上流社会建立的安全化城市空间,比如位于巴西的阿尔伐城。一些鸟屋高挂在树上,外形却犹如历史性建筑,例如墨索里尼政府所建立的意大利文明宫,象征着城市的绅士化(中产阶级化)现象,同时也减少了供动物使用的自然空间。

Bruce Gilchrist and Jo Joelson, who form the collective London Fieldworks, develop militant projects, linking nature and social context. In 2008, with "Super Kingdom", they denounce the palaces of dictators, as well as gated communities made for wealthy residents, such as the city of Alphaville in Brazil. Birdhouses in the form of historic architecture such as Mussolini's Palazzo della Civiltà Italiana are perched in the trees, symbolizing the urban phenomenon of gentrification, which reduces the natural spaces intended for animal species.

London Fieldworks
英国

超级国度 / Super Kingdom
Arts Council England, Henry Moore Foundation, Arts and Humanities Research Council and London South Bank University in collaboration with Consarc architects, Webb Yates Engineers and Setsquare Staging Limited
英国, 2008
pp.242-244

Photos ©: London Fieldworks

自2010年起,"自发城市"这犹如雕塑般的生态装置作品出现在伦敦的邓肯花园和克瑞姆花园,为鸟禽的生态多样性提供了有利的环境。"自发城市"由上百个供鸟类栖息的小屋组成,这些鸟屋高高架在臭椿树上,与附近1960年代社会住宅的佐治亚建筑风格产生呼应。

Since 2010, "Spontaneous City", an ecological sculptural installation, welcomes ornithological biodiversity in two London urban-green spaces: Duncan Terrace Gardens and Cremorne Gardens. Composed of hundreds of bird boxes, these installations, perched on trees of heaven, echo the surrounding architecture, both Georgian town houses as well as social housing of the 1960's.

自发城市 / Spontaneous City
Royal Borough of Kensington and Chelsea and Islington Council "Secret Garden Project"
英国, 2010-2011
p.245

蛇形画廊展馆
serpentine gallery pavilions
英国 United Kingdom
2006-2013

为了以象征性的方式替21世纪的开始留下印记，作为伦敦当代艺术展场的蛇形画廊建立了一项雄心勃勃的计划，每年邀请享誉国际的建筑师为画廊设计一座具前瞻性的展馆。

蛇形画廊位于肯辛顿公园内，每年崭新推出的临时性展馆会在夏季长达四个月的时间里，出现在画廊历史建筑附近的草坪上。这个前所未见的创意展示窗口，带着实验精神与艺术特质，开放给热情且好奇的广大民众。

Symbolically marking the beginning of the 21st century, the Serpentine Gallery, exposition space of contemporary art in London, undertakes the ambitious project of commissioning architects of international reknown to create a visionary pavillion each year.

During the four months of the summer season of each installation, a new folly is erected on the grass of Kensington Park, neighbour of the historic building that houses the Serpentine Gallery. A window onto creative innovation, because of its intentionally experimental character and its artistic dimension, is offered to a large public of amateurs.

蛇形画廊2012年展馆 / Serpentine Gallery Pavilion 2012
Herzog & de Meuron 瑞士
Ai Weiwei 中国
Photos ©: John Offenbach & Iwan Baan
p.246

蛇形画廊2010年展馆 / Serpentine Gallery Pavilion 2010
Jean Nouvel 法国
Photos ©: Philippe Ruault
上&下/top&bottom
John Offenbach 中/centre
p.247

这些具有绝然当代风格与设计语汇的小型建筑日夜都生气盎然，成为人们交流往来、进行社交与文化生活的场所，除了更新了公园与展馆的性格，更提供多种全新用途，而这正符合了城市空间与日俱增的需求。

This little structure, elaborated in a resolutely contemporary language, lives day and night. These spaces of conviviality, of social and cultural life, renew the genre and offer multiple new uses of which urban space has a growing need.

蛇形画廊2013年展馆 /
Serpentine Gallery Pavilion 2013
Sou Fujimoto 日本
Photos ©: Iwan Baan
p.248 & p.249 左上/top left
Jim Stephenson
p.249 右上/top right
Sou Fujimoto Architects
p.249 下/bottom

馥丽小筑 / 短暂　249

蛇形画廊2009年展馆 /
Serpentine Gallery Pavilion 2009

**Kazuyo Sejima and Ryue Nishizawa /
SANAA** 日本

Photos ©: Edmund Summer / VIEW
上&中/top&centre
James Newton / VIEW
下/bottom

p.250

蛇形画廊2006年展馆 /
Serpentine Gallery Pavilion 2006

Rem Koolhaas 荷兰
Cecil Balmond with Arup 英国

Photos ©: John Offenbach

p.251

捕梦园
dream-catching bubbles
法国 France

在法国东南部的五个天然景观基地内（分别位于阿洛、拉布亚迪塞、福卡尔基耶、蒙塔尼亚克、普吉），一些蚕茧般的奇特小天地不仅能让人徜徉在风景殊异的自然怀抱中，入夜后，繁星满天，更让人重新感受静夜的魅力。这些透明的"雪屋"是生态环保理念设计的结晶，透过静音风箱为聚氯乙烯材质充气而成形，同时能不断更新内部空气，而一切仅需消耗点亮55瓦灯泡的能量。"雪屋"的极简线条设计为人们提供180度的视野，邀请他们在此度过充满诗意与浪漫的亲密时刻，无论在任何季节，都沉浸于天人合一的境界里。

In five natural settings of southeastern France (Allauch, La Bouilladisse, Forcalquier, Montagnac, Puget), strange cocoons permit guests to immerse themselves in different landscapes and to rediscover starry nights. Environmentally friendly, these transparent igloos of PVC are inflated by silent bellows, continually renewing the air, all the while consuming the same energy as a 55 watt bulb. Offering a 180 degree view, their simple design encourages intimate, poetic, even romantic moments as well as communion with nature in all seasons.

Pierre Stéphane Dumas
法国

Photos ©: Attrap'Rêves

赫斯珀里得斯花园
garden of the hesperides
加拿大 Canada

2006

这是个长途旅行的邀约：沙地小径点缀着洁白的异国贝壳，穿梭在香根草与野鸢尾花遍生的草地中，引领人们一步步走过日式庭园风格的雪松炭踏石，抵达一座优雅的巨型灯笼。来自东方的设计灵感成就了这个橘黄色棉质帆布的建构物，大灯笼漂浮在暗沉如镜的水面上，而水中则静立着一株株苦橙树。

Representing invitations to a long-distance voyage, paths made of sand and of spotlessly white exotic shells cut through a prairie of vetiver grass and wild irises. The paths flow towards Japanese stairs of burnt cedar that themselves lead to an elegant, gigantic lantern. Made of saffron-coloured cotton canvas, this fabrique, of Asian inspiration, floats on a mirror of dark water where a grove of bitter orange trees grow.

Cao | Perrot
美国&法国
Festival International de Jardins de Métis, Québec
加拿大

Photos ©: Michel Laverdière p.256
Yvan Maltais p.257
Louis Tanguay p.258
Cao | Perrot p.259

256 follies / LUMINOUS

馥丽小筑 / 明亮　257

赫斯珀里得斯花园令人联想到同名的古老神话,除了东方文化的熏陶,同时也受魁北克墨提斯花园旁的河流景观所启发。漫步此花园中,仿佛置身梦境,空中飘散着阵阵甜香,交织着碘的气味,是一场微妙的感官体验之旅。

Inspired by the Orient, but also by the landscape of the Saint Lawrence river that borders the site of the Metis Gardens, this Garden of the Hesperides evokes the eponymous classical myth. As if immersed in a dream, the visitor has a delicately sensual experience, through mingled fragrances with a sugary and salty aroma.

馥丽小筑 / 明亮 259

艺术之家
a-art house
日本 *Japan*

2013

犬岛是日本濑户内海的一个小岛，一项具有艺术性的独特计划正在这个岛上展开。这项计划企图通过艺术与建筑来逐步更新犬岛的景观，并改变平均年龄为75岁的50多位居民对小岛的观感。

A-Art House艺术之家是本计划第二阶段首先展开的项目，由建筑师妹岛和世联合设计师周防贵之和艺术总监长谷川佑子共同设计。

On Inujima, a small island of Japan situated in the interior sea of Seto, a singular and artistic experience is happening. It aims to slowly renew the landscape through art and architecture, to refocus the vision shared by the little village's fifty or so inhabitants, the average age of whom is 75.

The A-Art House begins the second phase of the project, a collective work for which the architect Kazuyo Sejima brought together the designer Takashi Suo and the artistic director Yuko Hasegawa.

Kazuyo Sejima & Associates
日本
Fukutake Foundation
日本

Photos ©: Takashi Suo /
Kazuyo Sejima & Associates

他们选择在缓坡山谷的城市中心设置一个透明的圆形环道，超薄的双层亚克力墙面圈围住造型艺术家荒神明香为这个项目设计的群花帘幕。

走进这栋透明的"花房"时，彩绘的花朵图案层层叠叠，各种色彩斑斓的形状仿佛万花筒内呈现的影像，令人眼花缭乱。此装置影像再与周围的植物与矿质环境相叠合，创造出一种全新的景观。

They chose the urban heart of the gently sloping valley to install a very fine, double-walled ring of transparent acrylic that serves as a frame for the frieze of flowers of the multi-media artist Haruka Kojin, also associated with the installation.

When one enters the pavillion, the painted floral motifs, divided in two as if they were viewed in a kaleidoscope, disturb one's perception by their forms and colours. They superimpose themselves on the island's environment, made up of equal parts of plants and minerals, recreating a new landscape.

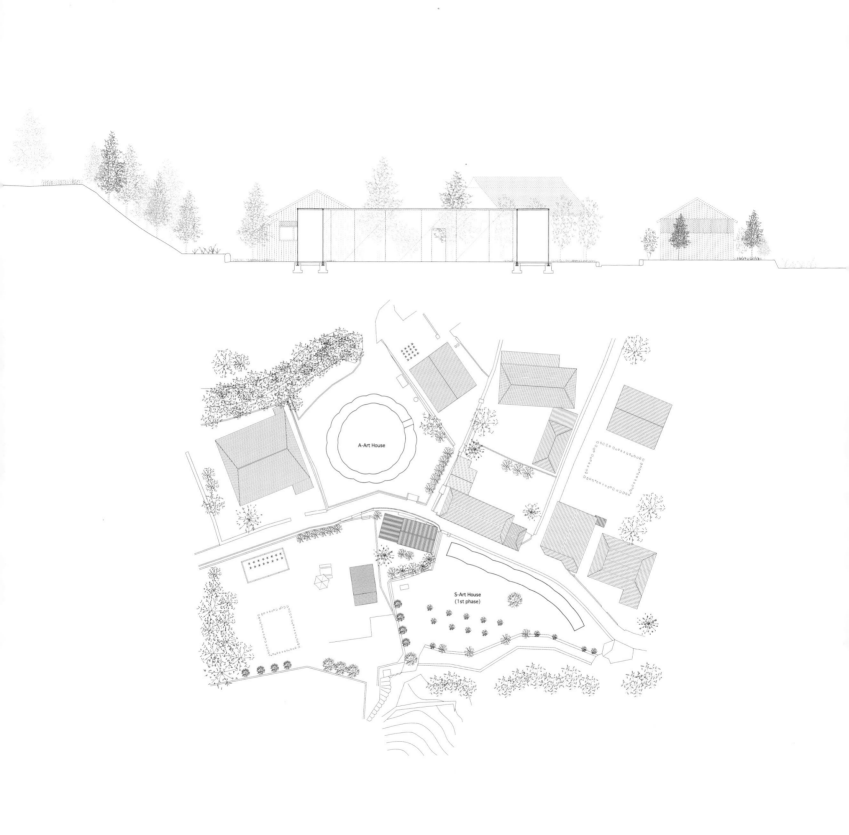

262 follies / LUMINOUS

这片人工合成草坪似乎从天而降到普罗旺斯地区艾克斯的一所大学的前广场上，让人们以另外一种方式看待基地的组织结构，以及它所衍生出来并被使用的穿行路线。这是对体育竞技的一种形象化描述，也暗喻着不同学习阶段的考试以及即将到来的现实生活中的竞争。

This patch of synthetic lawn seems to have landed on the parvis of the university, making the viewer look differently at the architectural configuration of the site and at the routes people habitually take. It is also a metaphor of competition, that which drives us to put study ahead of real life…

场地 / rest area - 2012

Laurent Perbos (法国)
L'Institut de Développement Artistique de l'Université de Provence, Faculté des Lettres et Sciences Humaines d'Aix-en-Provence (法国)

Photos ©: Laurent Perbos

05

实景艺术

LIFE-SIZE ART

实景艺术
life-size art

13号改编曲（高处栖息的猫）/
arrangement n°13 (perched cat)
2002
Jean-Luc Bichaud (法国)
Art Grandeur Nature / Département de la Seine-Saint-Denis (法国)
Photos ©:
Jean-Luc Bichaud & Dominique Gaessler

一条30米长的巨型水族箱在拉库尔讷沃公园小树林中的树木间穿行，这里成为金鱼们的度假之地。首先引人注意的是怪异的水声，透明的玻璃管悬挂在散步者头顶上，水则从管子的一端流向另一端。人们随而看到这件诗意的装置，设计师借此方式回应公园的人造景观，一种重新组构的自然。

One's attention is attracted to the strange sound of the movement of water flowing from one tube to another, each tube hanging over the visitor. Then one's gaze discovers the visual component of the poetic installation. A giant aquarium, 30 meters long, stretches from one tree to another in a grove, a veritable spa for goldfish in the middle of the Park of la Courneuve. Both the auditory and visual components of this installation echo the artificiality of the park, a piece of recomposed nature.

1993年，塞纳-圣-但尼省在拉库尔讷沃公园举办了一场当代艺术双年展，命名为"实景艺术"（Art Grandeur Nature），目的是在传统的展示场所之外向广大公众呈现法国以及世界各地艺术家的一些从未发表的艺术作品。

展览名称的决定并非出于偶然。主办单位谨慎地选择了它，以阐明这些在自然环境中设计与展示的作品的类型，这些作品是专为这块展览场地而创作的也必须与场地环境建立某种关联。此展览名称也与这座占地400公顷的景观公园产生共鸣，同时呼应了主办单位向受邀艺术家提出的创作主题：自然、人工、城市周边。

这个展览为始于1960年代末的新艺术实践划下象征性的标志，广为人知的大地艺术也正是从那个时期开始成为主导运动之一。但这些新艺术形式直到1980年代才真正地广泛开展，当时的艺术家们也越来越常离开画廊和博物馆这类安乐窝，而转向城市空间或者自然环境中以其他方式展示作品，并使他们的作品与另一群观众交会。

In 1993, the department of Seine-Saint-Denis creates a summer biennial of contemporary art in the Park of la Courneuve, entitled "Lifesize Art" (Art Grandeur Nature). The objective is to present to a large public new works of French and international artists, outside of the usual exposition spaces.

The choice of the name of this presentation owes nothing to chance. It is judiciously chosen to define the type of creations presented and conceived in situ, specifically for the place that will host them, and in connection with it. "Lifesize Art" also echos the site, a public landscape park of 400 hectares, as well as the themes proposed to the artists invited to participate: nature, artifice, and peri-urban environments.

This project is emblematic of new artistic practices initiated at the end of the 1960s of which Land Art is a pioneering movement. But this trend gained much greater amplitude from the 1980s on, when artists left more and more frequently the cozy nest of the gallery and the museum to display their work to other audiences, in nature as in the urban milieu.

此艺术实践大量利用不同的媒介以装置形式呈现作品，并且寻求改变人们对场地的官感和固有经验。观众甚至可以通过其创造出的互动关系转而成为作品的作者。

这种与户外空间的关联成为这些艺术实验的主导特性，它将带动造型艺术家以及景观师、建筑师、设计师等设计出不同尺度的作品。某些作品小巧精致，容易引起人们特别的注意，尤其在当今五花八门的社会中，眼睛总是目不暇给地看东西，经常忘了认真关注。

而一比一的真实尺度则可以使观众与作品产生一种"平等"关系，让他们以自身的尺寸来感受眼前的艺术作品。当观众在花园中体验到一种景观尺度甚至是土地尺度时，这些大型作品可以让人重新与广阔环境建立关系，也让他们从更集体的、更普遍的层面来思索环境问题。

The movement develops under the form of installations, where artists modify the perception of a site and thus the lived experience. The spectator can even become an actor by the interaction that the work proposes.

This relationship with the outside, significant in these experimentations, propels artists but also the landscapers, architects, and designers to conceive projects of various sizes.

Miniatures, on one hand, claim a particular attention even from our jaded modern eyes, 1:1 scale, on the other hand, puts spectators on par with the work and permits them an experience at their own measure. When they expand to the scale of landscape or territory, especially in the garden, the projects re-situate humans in their own enlarged environment and pose them questions at a collective, even universal level.

呵护自我形象，限制本我生活 /
trying to look good limits my life
2004

Stefan Sagmeister (奥地利&美国)
Art Grandeur Nature / Département de la Seine-Saint-Denis (法国)
Photos ©: Hervé Sellin & Matthias Ernstberger

大块的展示板上展示着设计师斯特凡·萨格梅斯特的作品：在亚利桑那州的沙漠景观中塑造并拍摄成照片的一些词汇。这些具有原创性的文字海报组成一个英语句子"Trying to look good limits my life（呵护自我形象，限制本我生活）"。人们对广告媒体中大肆宣扬的成就与美貌具有集体性的痴迷不悟，这件作品正是针对这种现象的讽刺与批评。

Sculpted and photographed words are posted on large panels in the desert landscape of Arizona, by the designer Stefan Sagmeister. His original typographies form a phrase "Trying to look good limits my life", ironic commentary on the collective drive for success and our obsession for the beauty celebrated in all types of publicity.

微观世界
miniature worlds
澳大利亚, 英国 Australia, United Kingdom

2001-2012

海伦·诺丁把她的一些"口袋景观"藏在城市空间的缝隙中。偏僻而隐秘的，在外部或内部的，被遗弃的角落甚至废墟，都成为她安置那些想像的"微观世界"的场所。那里隐藏着许多秘密小天地，等待人们细心发掘与探索，然后再自己虚构出一个故事。

Helen Nodding hides her pocketbook landscapes in the gaps of the city. Nooks and crannies of ruins, the exterior or the interior of these abandoned places, are the settings for her imaginative micro-architecture. With so many secret little worlds to discover, close attention can lead us to reinventing their history.

Helen Nodding
英国

躲藏 / **Hideaway**
英国伦敦, 2001
Photos ©: Helen Nodding
p.270 & p.271 下/bottom

围着营火 / **Around the campfire**
澳大利亚墨尔本, 2011
Photos ©: Jonathan Sander
p.271 上/top

秘密之门 / **Secret Door**
澳大利亚墨尔本, 2012
Photos ©: Helen Nodding
p.272

流行袭击 / **Meteor Attack**
澳大利亚墨尔本, 2011
Photos ©: Belinda Wiltshire
p.273

树 圈
tree bracelets
巴西 Brazil

2010

位于圣保罗伊的比拉普埃拉公园里的一些树干上缠绕着彩色的橡胶圈。树圈表面装饰着巴洛克风格的图案，人们可以通过镶嵌在其中的放大镜片，观察到树干上平时看不到的细节。这些可移动的放大镜同时也让人们得以探索植物微观世界里的其他元素。

Adorned with bracelets of colored latex, decorated with baroque motifs, the trees of Ibirapuera Park, in Sao Paolo, allow us to see, thanks to magnifying lenses, details usually invisible on their trunks. These magnifying lenses are also movable, allowing one to discover other elements of the vegetable microcosm.

Renata Mellão
巴西
Parque Ibirapuera
巴西

Photos ©: Renata Mellão
p.274 & p.275 上/top
Marcelo Uchoa p.275 下/bottom

仙境奇缘
wonderland
意大利 Italy

2004

为了庆祝罗马的荷兰学院创立100周年，也为了展示荷兰在各种形式的设计领域所具有的创新能力，荷兰政府选择了West 8景观事务所参与图拉真广场的项目。这个建造于公元2世纪的议事广场，是罗马的最后一个帝王建设。这件崭新的装置作品致力于为广场上的古市集带来视觉上的活力。

设置这件作品不能触碰任何古迹，当然更不能损坏它们。640平方米的平板结构通过打孔的方式嵌入到废墟之间，那些孔洞让人联想起雷达美奶酪的外观……平板之上铺展着只需要浇水和修整的青翠草坪，向人们展示了荷兰人的幽默以及简洁而又贴切的独创性。

To celebrate the 100th birthday of the Dutch Institute of Rome and to show Dutch creativity in all forms of design, the government chose the agency West 8 to intervene on Trajan's Forum, the last imperial site constructed in Rome in the 2nd century. The project's goal was to visually revitalize the ancient markets.

Because no ruins could be touched or damaged, a deep, sprawling padding of 640 m² was laid around the vestiges. The perforations in this protective layer recall the holes in Leerdammer cheese… Slightly raised and planted with a carpet of green grass, this symbolic landscape project, requiring only watering and mowing, shows the humor and the simple and pertinent originality of the Netherlands.

West 8 urban design & landscape architecture
荷兰
Netherlands Embassy in Rome
意大利

Photos ©: West 8 urban design & landscape architecture

多瓦珀植物
flora do vapor
葡萄牙 Portugal

2013

距离里斯本只有几千米的小渔村科瓦多瓦珀是一个反正统文化的活跃中心地。这件装置作品诞生于葡萄牙对法国团队Les Saprophytes（腐生植物）的邀请。

这里的居民极其关注海滨植物的保护，因此西尔维·达科斯塔设想出以七种本地沙丘植物为模板而塑造成的植物图谱。海芝麻菜、肾叶打碗花、海滨刺芹和其他植物的形态印记开始出现在堤岸边的岩石上以及邻近一个滑板坡道的边墙上，如同也正在进行中的当代项目"多瓦珀房子"所塑造出的城市图谱。

Situated in Cova do Vapor, a small fishing village some kilometres from Lisbon that became a very active hub of the counter-culture, this installation was born when the French collective Saprophytes was invited to Portugal.

Echoing the particular concern of the populus for the flora of the coastline, Sylvie Da Costa imagined a herbarium using stencils of 7 plants endemic to the dunes. The print of forms of *Cakile maritima, Calistegia soldanella, Eryngium maritimum, Pycnocomon rutifolium...* came to be integrated on the boulders of the dyke and on the neighbouring wall of a skateboard ramp. This urban herbarium is part of an alternative contemporary Casa do Vapor project, occuring in situ.

Sylvie Da Costa
法国
Les Saprophytes
法国
Casa do Vapor
葡萄牙
Association Ensaios e Diálogos
葡萄牙
Collectif Exyzt
法国
Cova do Vapor
葡萄牙

Photos ©: Sylvie Da Costa

2003-2006

倾斜 slope
德国 Germany

自然、童年和运动是造型艺术家玛丽·德尼的主要灵感源泉，她在其中汲取灵感以便为日常生活重新带来欣喜与活力，并使世界得到升华。她利用不同情境的互相冲击、奇特异常甚至不得体的形式，刻意创造出不合时宜的装置作品，如同这个设置于慕尼黑奥林匹克公园里小丘坡地上的足球场。

这个运动场由于它令人晕眩的坡度以及内部的小灌木丛，而无法真正被使用，反而成为一件颠覆景观的作品，也因为视觉角度的转化，而使得邻近的一座柱塔建筑摇身成为新的比萨斜塔。

Nature, sports, and the universe of childhood are the principal sources of inspiration of the multi-media artist Marie Denis, who draws upon them for re-enchanting daily life and enhancing the world. Her installations, intentionally anachronistic, play on the telescoping of situations, and on unusual, even incongruous forms, such as this soccer field traced on the side of a hill of Munich's Olympic Park.

Impracticable due to its dizzying steepness and the thickets that grow there, this sports area turned work of art tilts the landscape and transforms a nearby building into the new leaning Tower of Pisa!

Marie Denis
法国
M&M / Régine Goueffon & Thomas Huber / Impark
德国
Parc olympique de Munich
德国

Photos ©: Marie Denis

公园帆船座椅
sails park benches
加拿大 Canada

设计师费利克斯·居永被邀请设计一座纪念碑，用来向1670年创建韦谢尔村庄的居民致意，居永先生出生于此并生活于此，很自然地从这个位于蒙特利尔东北部的小村庄的历史中汲取灵感。

作为运输工具的船只曾经帮助居民们在圣劳伦斯河南岸安家，这件同时也是城市小品的装置作品便采用了几艘帆船并驾齐驱的形象，而以混凝土、金属和白橡木为材料的风帆则构成作品的主体，其所使用的木料在17世纪曾经是建造帆船和木桶的主要材料。

Invited to create a monument to render homage to the families who founded Verchères in 1670, the designer Félix Guyon, a native son who still lives there, was inspired by the history of this village situated northeast of Montreal.

Guyon evokes the boats which permitted the pioneers to establish themselves on the south bank of the Saint Lawrence. His work, a sort of street furniture, takes the form of majestic sails of concrete, metal and white oak: the same wood used in the 17th century for the construction of both sailboats and barrels.

Les Ateliers Guyon
加拿大

Photos ©: Félix Guyon

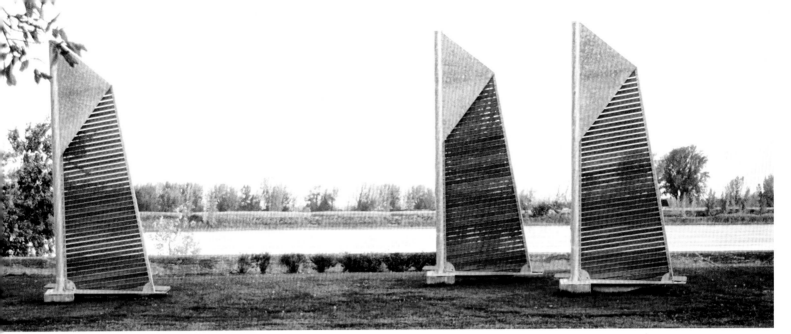

这件充满诗意的装置作品是与当地的工匠合作完成的，既呼应了基地中风与水这两个特征元素，也勾起了人们的集体记忆，同时也是居永先生对其祖父的纪念——身为木器工人的老祖父曾在1940年代建造"韦谢尔帆船"。

This poetic installation, produced with local artisans, recalls not only the specificity of this site between wind and water, and the collective memory of the town, but also that of the designer whose grandfather, a cabinet maker, made Verchères boats in the 1940s.

起舞
lets dance

1996-2013

德国, 比利时, 法国, 荷兰 Germany, Belgium, France, Netherlands

马克·德罗弗像种树一样栽种他的雕塑作品，使它们成为景观的一部分，随着时间自然地融入其中。他自诩为"农民雕塑家"，秉持着尊重和智慧展开"入侵自然"的行动。

这一连串带着韵律感重复出现的点状装置，以扭转、盘旋、运动、或虚或实的形态，创造出动感与平衡，并且和谐地融入环境之中。

它们以超越时间的姿态吸引人们的注意，似乎身体亦随之起舞。"无限循环"、"可见与不可见"、"展翅过道"……这些作品名称让人联想到大自然以及其永恒的起伏律动，如同海浪一般，永远都在重新开始……

Marc de Roover plants his sculptures. Tree-like, they are part of the landscape, where they integrate naturally, in time. A "peasant sculptor", he undertakes his invasion with respect and wisdom.

His rhythmic and repetitive constellations, based on twists, circles, and gestures, and on empty and full spaces, cultivate movement and balance, in harmony with their environment.

Timeless, the sculptures draw the eye as well as the body into their own dance. "Infinite Circle", "Visible-Invisible", "Path Through Fluttering Wings"... their names evoke nature and its ceaseless movements, like those of waves. The eternal begins again...

Marc de Roover
比利时

周期 - 无限循环 / Kringloop - Infinite Circle

Landgoed Anningahof, Zwolle
荷兰, 2005

Photos ©: Landgoed Anningahof
pp.284-285

可见与不可见 / **Visible-Invisible**
**L'Art dans les jardins,
Mont Saint-Vincent**
法国, 2009
p.286 上/top

可见与不可见 / **Visible-Invisible**
**Journées des plantes,
Domaine de Courson**
法国, 2006
p.286 下/bottom

展翅过道 /
Path Through Fluttering Wings
Hingene
比利时, 2013
p.287

Photos ©: Marc de Roover pp286-287

安宁加霍夫村艺术中心
landgoed anningahof art centre
荷兰 Netherlands
2004

这家私人艺术中心由希伯·安宁加在其家族原有的农业耕地上创建,位于阿姆斯特丹东北部的兹沃勒。出于对雕塑的热爱,他在一块大约6公顷的土地上同时开辟了一个现代艺术的展览场以及一座雕塑公园,剩余的大片空地仍然还保留其自然特色。

This private art centre was created by Hib Anninga, on his family's old farmlands in Zwolle in the northeast of Amsterdam. Passionate about sculpture, he developed both a space for the exhibition of contemporary art and a sculpture garden of about six hectares, a large part of which has been left natural.

公园之间 / Interpark
Gerard Groenewoud & Tilly Buij
荷兰, 2008
p.288

全景 / Panorama
Landgoed Anningahof
荷兰, 2010
pp.288-289

Photos ©: Landgoed Anningahof

此艺术中心在作品的选择上刻意采取兼容并蓄以及令人耳目一新的原则,并使作品与基地完美结合,随着季节的变迁而成为此场所的特征,为人们提供了一场融合自然与文化的漫步之旅。

The intentional and refreshing eclecticism of the artistic choices and their perfect integration into the site across the seasons characterize this place. It offers a promenade that spontaneously links nature and culture.

漂浮点 / Floating Point
Gerard Groenewoud & Tilly Buij
荷兰, 2012
p.290

受绞者 / Hanging Men
Tom Claassen
荷兰, 2003
p.291

Photos ©: Landgoed Anningahof

2009-2011
一万座桥的花园
the garden of 10 000 bridges
中国 China

很久很久以前存在着一座迷宫,它位于一片竹林的中心地带,其中小路纵横交错,时不时地就会把游人带到一座座传统的中国桥上,高耸俯视着周围景观,提供人们宽广的视野。这些时隐时现的红桥是找到迷宫出口的唯一参照点……

这座花园讲述着生命的故事,沙沙作响的禾本植物汇集成浓密而不可逾越的海洋,象征着人生中众多的飘泊和不确定性,同时,铺着深色卵石的狭窄小路有时会把人们带到"一万座"桥边,犹如阴霾中的一线青天以及美丽的惊喜,点缀着每一段人生际遇……

Once upon a time, there was a labyrinth at the heart of a bamboo forest, where roads cross and, from time to time, lead to traditional Chinese bridges. They overhang the surrounding landscape, opening to a spectacular view, the only fleeting glimpse of the path toward the exit.

This garden tells the story of life, made of a multitude of incertitudes and wanderings, symbolised by a sea dense with sonorous and impenetrable grasses and, at the same time, by the straight, dark gravelled path that sometimes leads to the "10 000" sunny spells and beautiful surprises sprinkling across each destiny...

West 8 urban design & landscape architecture & Atelier DYJG
荷兰&中国
Xi'an International Horticultural Exhibition
中国

Photos ©: West 8 urban design & landscape architecture

植物之家
casa botanica
摩洛哥 Morocco

2008

这片位于摩洛哥马拉喀什南部的苗圃占地7公顷，规模超乎寻常，苗圃的平面根据秘鲁亚马孙流域一个部落的编织中采用的几何图案而设计。通过对不同高度的精心组织，这个别出心裁的作品才得以实现，并让人联想起西班牙摩尔人的花园。

设立这个特殊的场所是为了普及一些不存在当地的植物，比如美丽异木棉、依拉瓦拉火焰树、洋紫荆或者是种类众多的多肉植物。这是摩洛哥第一个完全生态的植物园地，使用大蒜、黑皂、荨麻……来处理植物病虫问题，同时制造蚯蚓粪土。

The design of this extraordinary nursery in Morocco, south of Marrakech, unfolds over seven hectares and was inspired by a geometric design of a weaving of the Shipibo-Conibos, a tribe of the Peruvian Amazon. This original architecture takes form by playing on different levels, recalling Hispano-Moorish gardens…

This particular place was created to diffuse plants unavailable in the region, like the trees *Chorisia speciosa*, *Brachychiton acerifolius* and *Bauhinia x Blackeana*, or numerous varieties of succulents. It is an entirely ecological place, a first in Morocco, utlising garlic, black soap, nettles… to keep the plants healthy, while also producing its own vermicompost.

Sadek Tazi
摩洛哥

Photos ©: Sophie Barbaux

但是这个场所最吸引人之处是那些局部覆盖的丝网，这些丝网的设置是为了减少阳光对植被的灼伤，降低夏天干燥热风和冬天冷风的作用，而且还能保护植物不被冰雹伤害。不同的颜色突出了这个场所的神奇和诗意，设计师以图画般的方式组织光线，把整个苗圃转化为一个巨大的鸟笼。

But what is seductive about this place is the nets installed to soften the biting sun, to reduce the effects of the hot, dry winds in summer and icy ones in winter, and to protect the garden from hail. Their different colors accentuate the magic and poetry of the spot, transforming it into an immense sailboat, playing with the light in a pictorial manner.

波浪与光线
waves and lights
法国 France

2009

受到图卢兹西南部城市屈里奥的邀请,艺术家迪米特里·泽纳克斯在城堡公园设置了3 600平方米的大面积彩色纱网。它们稍稍高于地面,悬垂于不同树种之下。

通过这些用于收集橄榄树果实的纱网,泽纳克斯重新唤起了地中海文化的最古老仪式之一,以巧妙的几何方式组织这些条带状的纱网,构成纹章似的图案,并使它们悬浮起来。

At the invitation of the city of Cugnaux, southwest of Toulouse, the multi-media artist Dimitri Xenakis unfurled large nets over an area of 3 600 square meters. The nets, placed under different trees of the Manor Park, hang parallel to the ground in coloured waves.

With these meshes, meant to gather the olives that fall from the trees, Xenakis convokes one of the oldest rituals of the Mediterranean culture, representing it like a heraldic coat of arms.

Dimitri Xenakis
法国
Parc du Manoir de Cugnaux
法国

Photos ©: Dimitri Xenakis

某些条带拥有明亮绚丽的蓝色，仿佛从天上直接剥离下来，它们与象征夕阳的橙色条带形成鲜明的对比。这些设置在树木绿荫下的纱网，在人们的视觉范围内徜徉飘动。

通过在这样一个空间尺度中实现的景观，艺术家打破了景观园林历来风景如画的传统，在他所建立的秩序中，核心元素、前景、背景以人工的方式汇集在一起，在某一时刻，创造出一种新式的自然……

He plays with the geometry, organized into strips of color in levitation. Some bands bright blue, as if pulled down from the sky, contrast with orange ones in the image of the setting sun. Under the shelter of the trees, they stretch as far as the eye can see.

The artist has inverted this landscape to such an extent that he disturbs the picturesque tradition of the garden in its established order, with its central elements of fore- and background. He reunites the whole in an artificiality that creates, for a moment, a new nature…

澳大利亚花园
the australian garden
澳大利亚 Australia

这座21世纪的植物园位于距离墨尔本30千米的克蓝本，铺展在一个旧有的沙矿遗址上。花园的布局展现出这块次大陆所提供的多样性、反差以及超出常规的尺度。它也身负教学使命，让人们关注对现存植物以及它们所构成的生态系统的保护。

整个花园都展现出澳大利亚人如何在这块荒蛮而又神秘的土地上留下印记，殚精竭虑的程度令人惊奇。设计师以全然现代的形式，展现出自然景观与人工景观所营造出的不同氛围和丰富性。

Situated in Cranbourne, 30 km from Melbourne, this botanical garden of the 21st century unfolds in a former sand quarry. Designed to represent the diversity, contrasts and exceptional scale that the subcontinent offers, the garden also has the educational mission to raise awareness about conservation of the existing flora and its ecosystems.

The ensemble of the site shows how Australians have left their imprint on this wild and mysterious land, able to fascinate and to inspire fear. And under a resolutely contemporary form, the garden offers the richness of these different natural and domestic environments.

Taylor Cullity Lethlean & Paul Thompson
澳大利亚
Royal Botanic Gardens
澳大利亚

Photos ©:
John Gollings pp.302, 304-305, 307 中&下/centre&bottom
Peter Hyatt pp.303, 306
Ben Wrigley p.307 上/top

被刻意布置在中心并被浓妆艳抹的沙地花园以一片火焰般的铺地展现出澳大利亚大片的沙漠地带景观，一些多肉植物构成的抽象图案让人们联想起土著的绘画艺术。环绕在周围的则是艺术家格雷格·克拉克布置的海岸景观、点缀着桉树的稀树草原、热带森林、沼泽地以及可行船的水面……该地区与天空及特殊光线的关联也被搬上了舞台，为游人提供了一条奇特的观景路线，时而隐密时而广阔，却随处都充满了迷人的异国情调……

The sand garden, intentionally central and magnified, recalls the arid immensity of Australia, of the flaming land, punctuated with perennials, evoking aboriginal pictorial art. Around the central desert, unfolds the landscapes of the coast, with the Melaleucas spits, the native forests, the swamps, and the navigable zones with irregular water levels. It also includes scenographies by the artist Greg Clark. The relationship to the sky and the light so particular to this country is also recreated, offering an itinerary at times intimate and at others disorienting.

沙漠菜园
kitchen garden
蒙古 Mongolia

2012

蒙古正处于全面西式化发展的阶段，这个假菜园如同海市蜃楼般存在于戈壁滩干旱的土地上，是对此地严重环境问题的隐喻。

对于游牧民族来说，干燥的土壤并不适合新型农业的发展，而21世纪固定地点的生活方式——包括碳氢化合物的使用——则造成污染的存在，这片田园上的折叠塑料袋无声地谴责着这一切，虽然这些青菜不可食用，但仍可以在拆除此装置作品的时候进行"收割"。

This false vegetable garden installed like a mirage in an arid zone of the Gobi desert is a metaphor for the serious environmental problems of Mongolia, being developed in full-on Western-style.

The dry soil is ill-adapted to the nascent agriculture. The sedentary practices of this traditionally nomadic people in the 21st century, especially the use of hydrocarbons, have polluted the area. The dangers to the environment are brought to light in this field of folded plastic bags, inedible salad greens to be "harvested" without a doubt at the dismantling of this installation.

Maro Avrabou & Dimitri Xenakis
希腊&法国
Land Art Mongolia 360°
蒙古

Photos ©: Dimitri Xenakis

附 录
ANNEX

方案名单与联系方式
List of projects and contacts

柑橘的神奇力量 **Orange Power** p.2
Contact : archicolture@gmail.com
www.archicolture.com
www.festivaldejardins.cm-pontedelima.pt/ing

摇篮曲花园 **Lullaby Garden** p.4
Contact : andy@caoperrotstudio.com
xavier@caoperrotstudio.com
www.caoperrotstudio.com
www.cornerstonesonoma.com

前言
Preface

克拉姆溪迷宫 **Crum Creek Meander** pp.6-7
Contact : www.stacylevy.com
www.swarthmore.edu

二平方米的永恒 **2 m² of eternity** p.8
Contact : laurence.garfield@gmail.com
nat.houdebine@free.fr
www.2m2deternite.com

罗马尼亚山地聚落 **Rou** p.9
Contact : info@cigue.net
http://cigue.net

绿色岛屿 **Green Island** p.10
Contact : sylvie.h3@wanadoo.fr

虞美人 **Poppies** p.11
Contact : laurence.garfield@gmail.com
http://mosaiques-jardins.carbonmade.com
www.ccic-cerisy.asso.fr

01. 植物之歌
Vegetation

发辫艺术 **Twist in Cocody** p.12
Contact : marissima@free.fr
http://mariedenis.com
www.galeriealbertapane.com
www.jardins.nantes.fr
www.museedesbeauxarts.nantes.fr

迷宫 **Labyrinth** pp.14-15
Contact : fderanchin@free.fr
http://francederanchin.simdif.com

苏比亚科的椭圆形庭院
The Subiaco Oval Courtyard p.16
Contact : info@luigirosselli.com
http://luigirosselli.com/
william@dangargroupp.com
http://williamdangar.com.au

听…… **Listen...** p.17
Contact : olgaziemskastudio@gmail.com
www.olgaziemska.com
www.rzezba-oronsko.pl

韦尔奇山毛榉 **The Faux of Verzy** pp.18-19
Contact : contact@parc-montagnedereims.fr
www.parc-montagnedereims.fr
remi.collet@onf.fr
www.onf.fr

森林小径 **A Path in the Forest** pp.20-21
Contact : office@tetsuokondo.jp
http://tetsuokondo.jp

身体景观 **The Body Landscape** pp.22-23
Contact : marissima@free.fr
http://mariedenis.fr
www.galeriealbertapane.com
www.ciapiledevassiviere.com

过道 **Passage** pp.24-25
Contact : co.konrads@web.de
www.cokonrads.de

惊奇柴草堆 **Impressive Haystacks** pp.26-27
Contact : godderobin@gmail.com
www.robingodde.com
www.domaine-chaumont.fr

"球茎" 沃土 **Fertile Bulbs** pp.28-31
Contact : www.mesostudio.com
www.aafg.eu
www.les2cyclopes.fr
www.envers-du-jardin.com
www.domaine-chaumont.fr

植物装置 **Stick Work** pp.32-39
Contact : stickwork@earthlink.net
branchwork@earthlink.net
www.stickwork.net

普克树 **Pooktre** pp.40-41
Contact : becky@pooktre.com
http://pooktre.com

树木房客 **Tree Tenants** pp.42-43
Contact : archiv-office@harel.at
www.hundertwasser.at
www.hundertwasserfoundation.org

红地毯！**Red Carpet!** pp.44-46
Contact : contact@gaellevilledary.net
www.gaellevilledary.net

猫头鹰 **Owl** p.47
Contact : info@mosstika.com
www.mosstika.com

巨人头与泥土少女
The Giant's head and the Mud Maid pp.48-49
Contact : info@heligan.com
www.heligan.com

图尔坎墓园 **The Cemetery of Tulcan** pp.50-51
Contact : http://cementeriodetulcan.blogspot.fr
meckley@gmail.com
www.flickr.com/photos/meckleychina/5111932855/in/photostream

拉巴鲁城堡花园
The Gardens of the Château de la Ballue pp.52-57
Contact : chateau@la-ballue.com
www.laballuejardin.com

痕迹 **Tracks** p.58
Contact : archicolture@gmail.com
www.archicolture.com

绿色铸造 **Green Cast** pp.59-61
Contact : kuma@kkaa.co.jp,
kuma@ba2.so-net.ne.jp
http://kkaa.co.jp

摩根城之树 **The Morgantown Tree** pp.62-63
Contact : c.a.hummel@hotmail.com
www.carolhummel.com

疯狂之草 **Wild Grasses** p.64
Contact : alexandra.carron@gmail.com
www.alexandracarron.fr
www.alexandracarron.net

索朗日 **Solange** p.65
Contact : info@claudecormier.com
www.claudecormier.com

植物的婚礼 **Vegetal Wedding** p.66
Contact : fondation@ateliersdart.com
www.fondationateliersdart.org
contact@tzurigueta.com
www.tzurigueta.com
www.mnhn.fr

游旋塔玛 **The touring Tama** p.67
Contact : miriammcconnonart@gmail.com
http://miriammcconnonart.com

空中花园 **Aerial Garden** pp.68-69
Contact : andy@caoperrotstudio.com
xavier@caoperrotstudio.com
www.caoperrotstudio.com
www.laurent-perrier.com
www.jardinsjardin.com

苍鹭之树 **The Heron Tree** pp.70-73
Contact : www.lesmachines-nantes.fr

02. 协奏三曲
Accomplices

守护天使 **Guardian Angels** p.74
Contact : avrabou.xenakis@gmail.com
www.avrabou-xenakis.com
www.futuroscope.com

调色板 **Palette** pp.76-77
Contact : info@paulcocksedgestudio.com
www.paulcocksedgestudio.com
www.events.ukti.gov.uk/great-festival-of-creativity-hong-kong
www.londondesignfestival.com

呼吸盒子 **Breath Box** pp.78-79
Contact : contact@nasarchitecture.com
www.nasarchitecture.com
http://festivaldesarchitecturesvives.com

白色水源 **White Source** pp.80-81
Contact : mh.or@orange.fr
http://mhr-artinsitu.blogspot.fr
www.estuaire.info
www.cc-loiresillon.fr
www.abbaye-blanche-couronne.com

美之园 **Garden of Beauty** pp.82-83
Contact : jpp.delettre@wanadoo.fr
contact@oxalis-landscape.ch
www.oxalis-landscape.ch

爱丽丝与克拉拉 **Alice & Clara** p.84
Contact : marissima@free.fr
http://mariedenis.com
www.galeriealbertapane.com
www.pacbo.eu

红灯笼 **The Red Lantern** pp.85-87
Contact : andy@caoperrotstudio.com
xavier@caoperrotstudio.com
www.caoperrotstudio.com
www.cornerstonesonoma.com

漂浮森林 SS Ayrfield pp.88-89
Contact: andybrii@yahoo.com.au
www.flickr.com/photos/angeljim46/6199789236/
Steve Dorman: www.flickr.com/photos/60060337@N02/7642581940

维纳斯与偶然
The Game of Venus and Chance pp.90-93
Contact: avrabou.xenakis@gmail.com
www.avrabou-xenakis.com
www.domaine-chaumont.fr

漂浮岛 Floating Island pp.94-95
Contact: doug@douglasfitch.com
www.douglasfitch.com

自由荷叶 Lotus in Motion pp.96-97
Contact: ghalloran@dccnet.com
http://gordonhalloran.com

浮游植物 Phytoplankton p.98
Contact: artist@shigeko-hirakawa.com
http://shigeko-hirakawa.com
www.cdp29.fr/fr/presentation-trevarez

潮汐花 Tide Flowers p.99
Contact: www.stacylevy.com
www.hudsonriverpark.org

水上迷宫
Labyrinth on Water pp.100-101
Contact: fderanchin@free.fr
http://francederanchin.simdif.com
www.artotheque-angouleme.fr
www.mairie-angouleme.fr/ecole_arts

偏离常道
Off the Beaten Path pp.102-103
Contact: aureliebarbey@gmail.com
ruccolola@yahoo.fr
http://abarbey.ultra-book.com/book
http://lauraruccolo.wordpress.com
www.fetedesfeuilles.com
www.lyon.fr/lieu/lieux-danimation/parc-de-la-tete-dor.html

水上印记1 Water Footprint 1 pp.104-105
Contact: artist@shigeko-hirakawa.com
http://shigeko-hirakawa.com
www.cdp29.fr/fr/presentation-trevarez

某生存者心中的生活绝境
The Impossibility of Life in the Mind of Someone Living p.106
Contact: holger@beisitzer.de
www.beisitzer.de

南特<>圣纳泽尔河口
Nantes <> Saint-Nazaire Estuary pp.107-109
Contact: www.estuaire.info/fr
www.levoyageanantes.fr

薄雾雕塑 Standing Cloud pp.110-111
Contact: nakaya@processart.jp
www.processart.jp
duguet@club-internet.fr
www.domaine-chaumont.fr

白云 Bai Yun - White Cloud pp.112-113
Contact: andy@caoperrotstudio.com
xavier@caoperrotstudio.com
www.caoperrotstudio.com
www.cornerstonesonoma.com

回到源头 Return to the Source pp.114-115
Contact: aureliebarbey@gmail.com
ruccolola@yahoo.fr
http://abarbey.ultra-book.com/book
http://lauraruccolo.wordpress.com
http://festival-etangdarts.fr

交响田园 Harmonic Field pp.116-119
Contact: www.lieuxpublics.com

花朵2.0 Flowers 2.0 pp.120-123
Contact: esteve@pierreesteve.com
info@shooting-star.com
www.pierreesteve.com
www.facebook.com/esteveflowers2.0
www.detoursdebabel.fr

河岸 Riverine pp.124-125
Contact: www.stacylevy.com
www.mizu-tsuchi.jp

万有引力2号 Gravity #2 p.126
Contact: dimitri.xenakis@gmail.com
www.dimitri-xenakis.com

空间阴影 Spacial Shadows p.127
Contact: contact@thomasklug.com
www.thomasklug.com
www.chambourcy.fr/spipp.php?article340

欲望和威胁 Desire and Threat pp.128-129
Contact: www.cedricleborgne.com

镜屋 Mirror House pp.130-131
Contact: saracamre@mlrpp.dk
www.mlrpp.dk
www.stamerskontor.dk

眩晕 Vertigo p.132
Contact: contact@gaellevilledary.net
www.gaellevilledary.net
www.pays-valleeduloir.fr

明晰 Clear Cut p.133
Contact: info@kjellgrenkaminsky.se
maria@pollarkitektur.se
www.kjellgrenkaminsky.se
www.pollarkitektur.se

普赛克之镜 Psyche p.134
Contact: marissima@free.fr
http://mariedenis.com
www.galeriealbertapane.com
http://chamarande.essonne.fr

像素 Pixel p.135
Contact: info@travesiasdeluz.com
http://travesiasdeluz.com
http://segoviaculturahabitada.es

光陷阱 Trap Light p.136
Contact: arjen@transnatural.org
www.transnaturallabel.com/collection-shop

水母 Jellyfish p.137
Contact: paradedesign@yahoo.fr
www.parade-design.fr
www.festivallausannelumieres.ch

禁区 Zone pp.138-139
Contact: co.konrads@web.de
www.cokonrads.de
www.westwendischer-kunstverein.de

温室效应 Effect of Greenhouse pp.140-141
Contact: avrabou.xenakis@gmail.com
www.avrabou-xenakis.com
www.rogertator.com

漂浮之屋 The Floating Houses pp.142-143
Contact: www.lebalto.de
www.steirischerherbst.at

03. 居民造景
Inhabitant Landscapers

塞萨尔·曼里克基金会
César Manrique Foundation p.144
Contact: www.cesarmanrique.com

驻扎花园 The Inhabited Garden pp.146-148
Contact: www.fabuloserie.com

施华洛世奇水晶世界
Swarovski Crystal Worlds p.149
Contact: www.swarovski.com/kristallwelten

查尔特勒修道院花园
The Gardens of the Chartreux pp.150-151
Contact: www.chartreuse.org

乐土 Paradise pp.152-159

岩石花园 Rock Garden pp.160-165
Contact: www.rawvision.com
www.nekchand.com

罗伯·塔坦花园博物馆
The Museum Garden of Robert Tatin pp.166-171
Contact: museetatin@gmail.com
www.musee-robert-tatin.fr

荷花园 Lotusland pp.172-175
Contact: www.lotusland.org

陶花园 The Garden of the Pottery Flowers pp.178-185
Contact: contact@jardindepoterie.com
www.jardindepoterie.com

仙人掌花园 The Cactus Garden pp.186-195
Contact: www.cabildodelanzarote.com
www.centrosturisticos.com
www.turismolanzarote.com

展望小屋 Prospect Cottage pp.196-201
Contact: www.flickr.com/photos/johnsiddique

石园 Rock Garden pp.202-205

埃吉莱花园 The Garden of Éguilles pp.206-217
Contact: max.sauze@wanadoo.fr
anneleberre@wanadoo.fr
www.max-sauze.com

黄磨坊花园 The Garden of the Moulin Jaune pp.218-229
Contact: www.lemoulinjaune.com

04. 馥丽小筑
Follies

馥丽小筑狂想曲 Follies p.230
Contact: http://lavillette.com
www.tschumi.com

花园凉亭 Garden Pavilion pp.232-233
Contact: studio@paulraffstudio.com
http://paulraffstudio.com
http://artistsgarden.blogspot.fr

视听室A Auditorium A pp.234-235
Contact: claire.dehove@numericable.fr
un10.contact@gmail.com
http://wos-agencedeshypotheses.com

门外天地 Outlandia pp.236-237
Contact: www.londonfieldworks.com

蛇形树屋 Tree Snake Houses pp.238-239
Contact: info@rebelodeandrade.com
www.rebelodeandrade.com
info@pedrassalgadaspark.com
www.pedrassalgadaspark.com

生物拟态 Biomimicry pp.240-241
Contact: antony.r.gibbon@gmail.com
www.antonygibbondesigns.com

超级国度&自发城市
Super Kingdom & Spontaneous City pp.242-245
Contact: www.londonfieldworks.com

杰特尔广场位于圣但尼市的"巴黎之门"街区,广场上的每一个座椅都被写上了一种香料植物的名字,这个"文字播种法"也出现在圣但尼运河畔设置的具大木桌上。此艺术行动呼应了附近儿童的种植活动:他们在此公园以及他们的学校和游乐中心,实际栽种了这些既芳香又美味的香料植物。

In the neighbourhood "Porte de Paris" in Saint-Denis, the name of an aromatic plant is inscribed on each bench of Geyter Square. This typographic sowing is also present on giant tables created for the banks of the Saint-Denis Canal. These inscriptions echo the work of the children of the neighbourhood, who have planted in this park, around their school and activity centre the same tasty and aromatic plants.

香料植物的好地方 /
a good place for herbs - 2012

Colas Baillieul & Sophie Barbaux,
Les Rudologistes Associés (法国)
Synesthésie / Saint-Denis (法国)
Imaginaire et Jardin / Plaine Commune (法国)
Photos ©: Sophie Barbaux

pp.310, 312, 314, 316, 318

蛇形画廊展馆
Serpentine Galleries Pavilions pp.246-251
Contact: www.serpentinegalleries.org
www.sou-fujimoto.net
www.oma.eu
www.balmondstudio.com
www.arupp.com
www.herzogdemeuron.com
http://aiweiwei.com
www.jeannouvel.com
www.sanaa.co.jp

捕梦园 **Dream-Catching Bubbles** pp.252-255
Contact: contact@attrap-reves.com
www.attrap-reves.com

赫斯珀里得斯花园
Garden of the Hesperides pp.256-259
Contact: andy@caoperrotstudio.com
xavier@caoperrotstudio.com
www.caoperrotstudio.com
www.jardinsdemetis.com

艺术之家 **A-Art House** pp.260-263
Contact: www.sanaa.co.jp
www.benesse-artsite.jp/en/inujima
www.fukutake.or.jp/art

05. 实景艺术
Life-Size Arts

场地 **Rest Area** p.264
Contact: superbos@hotmail.com
http://www.documentsdartistes.org/artistes/perbos/repro1.html
http://allsh.univ-amu.fr

13号改编曲（高处栖息的猫）
Arrangement n°13, (Perched cat) pp.266-267
Contact: www.jeanlucbichaud.fr
www.seine-saint-denis.fr

呵护自我形象，限制本我生活
Trying To Look Good Limits My Life pp.268-269
Contact: info@sagmeisterwalsh.com
www.sagmeisterwalsh.com
www.seine-saint-denis.fr

微观世界 **Miniature Worlds** pp.270-273
Contact: www.helennodding.com

树圈 **Tree Bracelets** pp.274-275
Contact: solange.viana@uol.com.br
www.mcb.org.br

仙境奇缘 **Wonderland** pp.276-277
Contact: west8@west8.com
www.west8.com
www.knir.it
www.capitolium.org/eng/fori/traiano.htm

多瓦珀植物 **Flora do Vapor** pp.278-279
Contact: www.mala101.net
www.les-saprophytes.org
www.casadovapor.org
www.facebook.com/collectif.exyzt

倾斜 **Slope** pp.280-281
Contact: marissima@free.fr
http://mariedenis.com
www.galeriealbertapane.com

公园帆船座椅 **Sails Park Benches** pp.282-283
Contact: www.lesateliers-guyon.com
www.ville.vercheres.qc.ca

起舞 **Lets Dance** pp.284-287
Contact: marc.d.roover@gmail.com

安宁加霍夫村艺术中心
Landgoed Anningahof Art Centre pp.288-291
Contact: info@anningahof.nl
www.anningahof.nl

一万座桥的花园
The Garden of 10 000 Bridges pp.292-293
Contact: west8@west8.com
www.west8.com

植物之家 **Casa Botanica** pp.294-297
Contact: sadektazi@gmail.com
www.facebook.com/www.casabotanica.ma/posts/480671265388908

波浪与光线 **Waves and Lights** pp.299-301
Contact: dimitri.xenakis@gmail.com
www.dimitri-xenakis.com

澳大利亚花园 **The Australian Garden** pp.302-307
Contact: www.tcl.net.au

沙漠菜园 **Kitchen Garden** pp.308-309
Contact: avrabou.xenakis@gmail.com
www.avrabou-xenakis.com
http://landartmongolia.blogspot.fr

附录
Annex

香料植物的好地方
A Good Place for Herbs pp.310-318
Contact: s.barbaux@orange.fr
colas.baillieul@gmail.com
www.sophie-barbaux.odexpo.com
www.synesthesie.com
imaginaireetjardin.blogspot.com

致谢
Acknowledgement

我们诚挚感谢本书所有刊登方案的设计者，谢谢他们提供了迷人的创作和友善的协助。

作者在此特别向以下人士致意，他们在本书编著作过程中所给与了诸多中肯的想法、建议和支持：
Caroline Barbaux, Cécile Beiso, Andy Brill, Catherine Dan, Steve Dorman, Catherine Feuillie, Laurence Garfield, Isabelle & Jacques Glowinski, Sara Lubtchansky, Jean-Jacques Mandel, John Meckley, Paoli Rinaldi, Kitshette Rodrigo, John Siddique & Hervé Thibault.

作者也感谢法国亦西文化公司的简嘉玲(Chia-Ling Chien)和尼古拉·布里左(Nicolas Brizault)长久以来的信任，使这第四本合作的图书能够圆满出版。

Our warmest thanks to all the designers of the projects presented, for their creations and their kind cooperation.

The author thanks, for their suggestions, their exchange of ideas or their collaborative participation a propos of this work:
Caroline Barbaux, Cécile Beiso, Andy Brill, Catherine Dan, Steve Dorman, Catherine Feuillie, Laurence Garfield, Isabelle and Jacques Glowinski, Sara Lubtchansky, Jean-Jacques Mandel, John Meckley, Paoli Rinaldi and his blog Carnet de Notes, Kitshette Rodrigo, John Siddique and Hervé Thibault.

And for their unfailing trust: Chia-Ling Chien and Nicolas Brizault.

图书在版编目（CIP）数据

奇异花园/(法)巴尔波编著；邵雪梅，王美文汉译；(美)薛帕尔德英译.--沈阳：辽宁科学技术出版社，2015.5
ISBN 978-7-5381-9225-4

Ⅰ. ①奇… Ⅱ. ①巴… ②邵…③王…④薛…Ⅲ. ①景观设计 - 作品集－世界－现代 Ⅳ. ①TU986
中国版本图书馆CIP数据核字(2015)第085676号

出版发行：辽宁科学技术出版社
　　（地址：沈阳市和平区十一纬路29号 邮编：110003）
印　刷　者：利丰雅高印刷（深圳）有限公司
经　销　者：各地新华书店
幅面尺寸：240mm×260mm
印　　张：20
插　　页：4
字　　数：50千字
出版时间：2015年 5 月第 1 版
印刷时间：2015年 5 月第 1 次印刷
责任编辑：宋丹丹
封面设计：卡琳·德拉梅宗
版式设计：卡琳·德拉梅宗
责任校对：周　文

书　　号：ISBN 978-7-5381-9225-4
定　　价：358.00元

联系电话：024-23284360
邮购热线：024-23284502
http://www.lnkj.com.cn